Working across Lines

The publisher and the University of California Press Foundation gratefully acknowledge the generous support of the Ralph and Shirley Shapiro Endowment Fund in Environmental Studies.

Working across Lines

Resisting Extreme Energy Extraction

Corrie Grosse

UNIVERSITY OF CALIFORNIA PRESS

University of California Press
Oakland, California

© 2022 by Corrie Grosse

Library of Congress Cataloging-in-Publication Data

Names: Grosse, Corrie, 1990– author.
Title: Working across lines : resisting extreme energy extraction /
  Corrie Grosse.
Description: Oakland, California : University of California Press,
  [2022] | Includes bibliographical references and index.
Identifiers: LCCN 2021061636 (print) | LCCN 2021061637 (ebook) |
  ISBN 9780520388406 (cloth) | ISBN 9780520388413 (paperback) |
  ISBN 9780520388420 (epub)
Subjects: LCSH: Energy conservation—Political aspects—California. |
  Energy conservation—Political aspects—Idaho. | Environmental
  justice—California. | Environmental justice—Idaho. | Political
  participation—California. | Political participation—Idaho. |
  Coalitions—Case studies. | BISAC: NATURE / Environmental
  Conservation & Protection | SCIENCE / Global Warming &
  Climate Change
Classification: LCC HD9502.U63 C24 2022 (print) | LCC HD9502.U63
  (ebook) | DDC 333.7909794—dc23/eng/20220128
LC record available at https://lccn.loc.gov/2021061636
LC ebook record available at https://lccn.loc.gov/2021061637

Manufactured in the United States of America

31 30 29 28 27 26 25 24 23 22
10 9 8 7 6 5 4 3 2 1

*For my interviewees and everyone working
for climate justice*

Contents

Preface

The fourteen years from 2007 to 2021 were full of crises and mobilization. December 2007 marked the beginning of a global Great Recession. In 2009, hopes for a global climate treaty and US climate legislation were dashed while bailouts to the banks behind the Great Recession rolled out. In 2011, the Occupy Movement and Arab Spring mobilized to contest injustice and inequality. Since 2013, Black Lives Matter and the climate justice movement have been gaining momentum. Throughout this period, unconventional fossil fuel extraction in the form of hydraulic fracturing boomed in the United States, making it the largest oil and gas producer in the world, and developers of the Canadian tar sands met unexpected resistance as they tried to get their molasses-like oil to market. In 2016, people experienced the hottest year ever recorded (tied with 2020), thousands of people and hundreds of tribes gathered in solidarity with the Standing Rock Sioux Tribe to resist the Dakota Access Pipeline, and nearly sixty-three million people voted to elect Donald Trump, someone who expressed skepticism about climate change and made derogatory statements about women, people of color, immigrants, and Muslims, among others.[1] In 2020, more than three million people died from COVID-19 globally, greenhouse gas emissions dipped, and 51 percent of Americans (eighty-one million) voted for Joe Biden, the first US president to make climate crisis and racial justice core pillars of his campaign.

As a businessman whose actions as president simultaneously damaged the lives of people and the planet, Trump exemplified how social (in)justice, climate (in)justice, and capitalism are intimately intertwined.[2] Understanding of this interconnectedness is also visible within the ranks of people who are resisting oppression and creating more just and sustainable visions for the world, visions that they enact in daily life.

The people whose stories I share in this book are resisting the energy systems that fuel their lives while governments, industry, and mainstream culture hold tight to a political economy that is destroying the planet. The central actors in this project are not stereotypical environmentalists. Most are not far-left political radicals. They are women, conservatives, young, old, and Indigenous. They are small business owners, conservationists, families, and, yes, some are liberals and some are environmentalists—people who work for environmental organizations or those who, partially out of their commitment to the cause, refuse typical jobs or homes, taking solace in or from nature. They fight because they are directly affected by extraction, or know they will be soon. Their home valuations, land, health, water, air, and ideas about representative democracy are under attack by the oil and gas industry. They notice the decline of the fish and plants that their people have depended on for centuries. But others who fight are not affected. They learned about climate change through an eye-opening film, moving news bulletin, or terrifying article. These media, in the context of people's lives as students, parents, or elders, kindled urgency in their hearts, spurring them to act on what is still, for many of them, an abstract crisis.

In short, resistance to extreme extraction is far from homogenous. Its form is not that of a smooth obsidian arrow with a pinpoint tip. Rather, it is a shower of raindrops on untreated steel, slowly pounding away to create rust, eventually sculpting holes through which rain-nourished grass can sprout. The raindrops fall all over and sometimes in the same spot, where the steel is most weak or dependent for support. Unlike the arrowhead that, when fractured, splinters into impossible-to-put-back-together shards, the pool of water collecting in the steel's low point, or on the adjacent ground, is fluid. It can separate into individual drops and regroup without losing the character and strength of the whole or its parts. Both drops and pools provide hydration, though in different quantities. They carve canyons, if at different rates.

This book is about these raindrops of resistance and how they coalesce.

Acknowledgments

This book would not exist without the activists who inspired me to take on this research and who continually make me believe that we can change the world. I dedicate it to you.

In Santa Barbara, Becca Claassen provided the positivity, cheer, and unwavering dedication that made me feel motivated to be a core member of 350 Santa Barbara and participant in Measure P. Thanks for bringing us all together, Becca. The 350 community was a special one for me because of its original members: Becca, Max Golding, Sharon and John Broberg, Gary Paudler, Alex Favacho, and Katie Davis. You all made me feel connected to Santa Barbara and the climate justice movement, teaching me what it means to have an amazingly powerful and fulfilling grassroots group. I also want to thank Arlo, my UCSB cocaptain for the signature phase of Measure P, without whom I surely would have had a breakdown from the stress of it all. Emily Williams and Theo Lequesne, I am awed by your skills and insights into effective organizing and feel so lucky to be able to turn to you for advice.

In Idaho, I am indebted to Helen Yost, Alma Hasse, and Jim Plucinski. Helen, thank you for making my first experience in environmental activism such a positive one and for keeping me attuned to climate justice organizing in the Pacific Northwest since 2011. I look to your sincerity and rejection of capitalism as a model for social relations with the more-than-human world. Alma and Jim, words cannot describe how fortunate I feel to have lived with you. Thank you for teaching me how to listen to different perspectives and find common ground, for your generosity, and for throwing so much of yourselves into protecting the places we love. Sherry, thank you for your warmth, attention to details, and for keeping me updated since 2015. Shelley, I admire your dedication and am grateful for all you do to safeguard communities. Dottie, thank you for your work and for always feeding and welcoming us all into your cozy home. To Lee, many thanks for your friendship and hospitality.

Thanks to the many more unnamed individuals who gave me so much of your time during interviews and fieldwork. You taught me a lot about activism, hope, and life. I am grateful to Helen, Alma, Leontina Hormel, Julian Matthews, Gary Macfarlane, and Becca for connecting me with interviewees. Thank you to the University of California Santa Barbara (UCSB) Graduate Division, the UCSB Chancellor's Sustainability Committee, the Flacks Fund for the Study of Democratic Possibilities, UCSB Crossroads, the UCSB Department of Sociology, Kum-Kum Bhavnani, the UCSB Academic Senate, the UCSB Graduate Student Association, and the UCSB Sociology Graduate Student Association for supporting this project.

I am indebted to the truly exceptional mentors I have had in my education and life. Kum-Kum, thank you for always pushing me to critically examine difference, center women's lived experiences, and pursue my passions, no matter how challenging they seem. Without your nudge to do fieldwork in Idaho, I would not have become aware of working across lines. Your expertise in setting timelines and developing research methods has been key to my success. To John Foran—wow! Thank you for making graduate school a joyful experience; for modeling what it takes to be a teacher who inspires students to change the world; for opening so many doors; for connecting me with 350 Santa Barbara and some of my closest friends; for showing us what scholar activism can be; and for helping us all remember to hold love, joy, and hope in our hearts as we confront the climate crisis.

Thank you to ann-elise lewallen, whose enthusiasm for my ideas and those of my research participants, insightful comments on my work, and advice for practicing accountability while in the field added depth and richness to my book. To David N. Pellow, who provides me with some of the most useful and grounded advice for how to excel as a professional sociologist while maintaining a radical commitment to justice, activism, and humility. To Leontina Hormel, who in her commitment to students and employing research to address inequality, first inspired me to become a professor. LT, thanks for being a steadfast source of support these many years. To Melanie Neuilly, who taught me how to do research and find funding, and inspired me to focus so much of my learning on gender. Thank you for being my friend and providing me with a home base in Moscow. To Shannon Bell, whose research with women who work for environmental justice in Appalachia inspired my own and whose comments improved the book. Receiving a positive review from you meant the world to me! Thank you for your insight on publishers and sharing your book proposal with me years ago.

I would not have excelled in this project without the encouragement of my friends. Katelynn Bishop, Summer Gray, and Heather Hurwitz all provided extensive moral support and insight that helped me develop and refine my ideas for this book. Heather, thanks for being a role model on how to publish a book and sharing your materials with me!

In the final stages, I benefited tremendously from the editing and insight of Jenny Kutter. Thank you for your thoughtful comments and ideas for effective framing and accessibility. To Valerie Doze, thank you for your incredible proofreading! To Will Matuska, for your patience and skill as a cartographer, and Rachel Leen, for your help with details. Thank you to Melissa Burrell and Brigid Mark for your encouragement and positivity. To an anonymous reviewer, I am thankful for your encouragement and important revisions. Thank you to my editor and the support staff at University of California Press for your interest in my work and detailed answers to my questions.

Finally, I am grateful for my family, who have always supported and encouraged me to do my best. To Kathy and Mike Shidner and Virginia and Clint Grosse, for your love and support. To Ali Meiners, whose interest in and powerful enthusiasm for my well-being and work always gives me an extra boost. To my mom and dad, Ruth and Gary Ellis, who are steadfast in their love and support. Thanks for giving me the values, skills, resources, and encouragement to realize a life I love. You are the best listeners. Finally, to my best friend and partner, Hunter Grosse. Thanks for making me smile and for building a home with me where I feel loved, confident, and happy. Thanks for your patience, advice, and help on this project and for caring for Farren. And to Farren, thank you for smiles, hugs, and dance moves.

Introduction

It was December 12, 2015. I was nearing the end of a two-week stay with Alma Hasse and Jim Plucinski, two small business owners who live on a farm near Parma, Idaho.[1] For the last five years, Alma had been working to stop natural gas development in Idaho. It was a new industry for the state. Jim had begun investing all of his time, outside of running their business, to challenge natural gas development after Alma spent seven days in jail when she spoke out of turn at a county planning and zoning meeting related to oil and gas in October 2014. The theme we were discussing was: How might we understand the oil and gas industry's practices and what can we do about them? Jim had a simple answer. The oil and gas industry, he explained, is like a dog-walking business that doesn't want to pick up after the dogs:

> One analogy would be, you know, somebody that's got a dog-walking business. And they have maybe fifteen employees and each one of 'em's got certain dogs they walk every single day, but the problem they're having is lack of productivity from them having to stop every so often and pick up a dog's mess.
>
> Well, most rational businesspeople will, number one, realize that that [the time spent picking up the mess] is just part of the job and it's a good thing to do to clean up after yourself or your employees' dogs. But unfortunately, today's business climate, especially with large corporations, is that they don't think about it that way. They think about, "How can we get rid of this?" And it's usually through legislation, [. . .] lobbying everybody in the state or federal government to say, "You know what? [. . .] We're really hurting the community by not having more jobs. We can have more people if we

don't have to pick up dog poop, because we will make more money, we could hire more people, we could even pay them more"—not that they are going to do that, but it sounds good. "So, we would like to implement a law that we don't have to pick up that poop anymore. You know, we could actually probably even get a couple more dogs in a day if we don't have to pick up the poop on the walk, we can make more money." But that's behind the scenes.

That's what they do, they create laws to do things that other people are not allowed to do, or conscientiously would not do, and make it legal. And then when they walk that dog and the dog does his thing right on the neighbor's lawn [. . .] and the neighbor says, "Aren't you going to pick that up?" they go, "No ma'am, we're following the laws. We don't have to pick that up; we're doing everything regulated; we're doing everything by the books; we're following the laws." And they walk off with a smile. [. . .] Guess who's going to pay to pick up that poop? It's going to be everybody else that lives in that neighborhood. [. . .] They're taking time away from their family, and you know, utilizing their resources to do so.

"To deal with somebody else's shit! That is a perfect analogy!" Alma broke in laughing. Jim went on: "Exactly. [. . .] It's not about oil and gas, people have to realize that oil and gas is just a symptom of the problem. There are many, many different business entities and types that are doing the exact same thing."

Jim then moved on to talk about how to confront this problem and his assessment of how people working on resistance could be more effective:

You have to look at it as a tree. Everybody is working on different branches and the bottom line is, I mean—can you tell me if you cut a branch off a tree, will that tree die? [pause] What's it going to do? It's going to put up another branch. [. . .] So the big thing is, is that everybody, as far as people that are activists, [. . .] everybody is working on a branch and not on the roots. [. . .] People need to start realizing what their true enemy is, and that's the collusion between unethical business and corrupt government. [. . .] They are so busy keeping us separated between a right thing [on the political spectrum], and a left thing.

Jim's dog-walking analogy illustrates a common sentiment among people resisting fossil fuels: oil and gas development is an affront to core values of integrity, accountability, fairness, and the health and well-being of families and communities. In the minds of Jim and other people I interviewed throughout the course of my research, the oil and gas industry walks all over people, uses legal and political systems to protect itself, and leaves communities to pick up the messes it produces. These messes are toxic environments and an inhospitable climate; sick people; degraded social ties; and losses of homes, livelihoods, and public resources. For Jim, the collusion and corruption of the oil and gas industry represented a lack of care and respect, a lack of integrity and accountability that he, as a business owner, saw as common decency. He and Alma did not just believe in these values. They also acted on them, treating the people around them, their employees, customers, family, and activists, with care, respect, and generosity.[2]

Different people have different terms for these values, but at bottom, they revolve around issues of fairness, right and wrong, relationships, and justice. From 2014 to 2016, I set out to understand how people resisting fossil fuels in two very different settings—Idaho and California—understood and used their core values to work together.

At the dawn of the 2020s, it seems harder than ever to work together. In the United States, politics and ideas for addressing environmental problems and climate crisis are highly polarized. On one side are people benefitting from the current system—capitalism—who hope to continue fossil fuel extraction and enjoy its profits. On the other side are those who do not benefit from existing systems: communities on the frontlines of environmental and climate injustice. The lessons from this book, on how to work together across dividing lines within communities—how to identify shared values, acknowledge and value difference, and grow—are more critical than ever.[3]

Despite the social nature of our failure to address climate crisis— we have not figured out how to enact laws, policies, and behavior change on a large enough scale to stop greenhouse gas emissions—most research and policymaking on energy and climate change focuses on technology and physical science (Dunlap and Brulle 2015; Sovacool 2014). Understanding the social dynamics—what people do with the science and technology, and the justice elements, the inequality built into energy systems and the climate crisis—is vital (Caniglia, Brulle, and Szasz 2015; Harlan et al. 2015).

As scholars of climate justice maintain, building a broad-based social movement that centers the voices of people on the frontlines of environmental and climate hazards is critical for building political will to make our social, economic, and cultural systems less carbon intensive and, simultaneously, more socially just. Surprisingly, researchers have devoted relatively little attention to understanding how social movement coalitions work (Van Dyke and McCammon 2010). Most research that does address coalitions tends to focus on organizations, cause and effect, and single variables, such as how a threat or opportunity shaped a coalition (see Van Dyke and McCammon 2010). This book provides a more holistic analysis of how culture interacts with many factors that shape movements (identity, political context, threats, and resources) to inform how activists—as individuals and within organizations—*practice* coalition building. I think of culture as the "lived experience" (Williams 1960) of activism and am interested in identifying best practices for coalition building that can aid in larger scale "movement building" efforts (Juris et al. 2014).

What we do know is that a lack of collective identity—a sense of shared understanding and vision, a sense of togetherness (see Melucci 1989; Taylor and Whittier 1992)—can be a barrier to movement building. In her research on resistance to mountaintop-removal coal mining, Bell (2016) finds that local Appalachians' inability to identify with the environmental justice movement keeps them from participating in the movement. This is a problem for a movement that focuses on meaningful involvement of those most affected by environmental hazards. While Bell asks why people do not participate, my research asks *how* people *do* participate. How do they work to create inclusive collective identities?

Drawing from the experience and insights of diverse activists—Nez Perce tribal members in Idaho, a Chumash family in Santa Barbara, college students, elderly people, women, men, people with children, working people, unemployed people, wealthy people, poor people, people with disabilities, people of color, mixed-race people, and white people—this book is a study of how people work together by appealing to common values.

Chapters 3 through 8 of the book explore my interview- and fieldwork-based data. These chapters flow from in-depth analyses of practices and perspectives of particular groups of interviewees to comparative analyses that interweave stories of campaigns and perspectives across Idaho- and California-based groups and research sites. They reveal that resistance to extreme energy extraction is characterized by *working across lines*, a phrase I use to refer to activists' efforts to organize people across lines of difference, whether these be lines based on political views, race, ethnicity, indigeneity, age, area of interest, strategic and tactical preferences, or type of organization (i.e., staffed nonprofits that I call "grasstops" groups versus grassroots groups). I identify four major components to working across lines as a method of resistance to extreme energy extraction:

- focusing on core values, which include community, justice, integrity, accountability, and the health of people and the more-than-human world
- identifying the roots of injustice, whether described as capitalism or lack of integrity and accountability of government and industry
- cultivating relationships, which some interviewees refer to as *relational organizing*
- welcoming difference

Through prioritizing perspectives and action to realize these elements of organizing, activists and groups across my research sites build capacity to construct unlikely alliances and coalitions to challenge the fossil fuel industry; they work across lines for a just and sustainable future. When activists agree on core values, illuminate how unjust conditions put these in jeopardy, and draw on relationships of trust to welcome and support diverse participants and tactics, they are better equipped to create a truly inclusive collective identity. Building an inclusive collective identity through working across lines has potential to grow a broad-based social movement, one that could be society's best hope for achieving climate justice.

GUIDING IDEAS: USEFUL FRAMEWORKS FOR WORKING TOGETHER

Environmental justice, climate justice, intersectionality, and ecofeminism are four frameworks for organizing around environmental and social justice issues that can contribute to building broad-based collective identities. These frameworks are evident in the best examples of working across lines from this research and can enhance capacity for working across lines. In addition to these frameworks, this section unpacks movement building, collective identity, and coalition building, three key processes and goals of the social movements that I analyze in this book.

Environmental justice emphasizes "meaningful involvement of all people" in all phases of policy creation and implementation (US EPA 2017). If people are meaningfully involved in decisions about social movements, they will be more likely to feel a sense of ownership of the movement, that their personal identities align with what the movement stands for.

Climate justice applies the necessity for meaningful involvement to the context of climate change, highlighting how those who are least responsible for the climate crisis are most affected. It seeks to address climate crisis through advancing social justice. I see the movements and activism that I explore in this book as contributing to climate justice, as part of the climate justice movement. The focus on climate crisis and social justice as global phenomena opens the possibility for people around the world to feel connected to this movement.

Intersectionality is the idea that the different social identities individuals hold come together to shape their experiences of oppression and privilege (Crenshaw 1989). Highlighting the intersectional identities of movement participants can help everyone feel welcome, as when, for example, the climate organization 350.org made a public statement of solidarity with the LGBTQIA+ community following the 2016 mass shooting in Orlando, Florida (350.org Staff 2016). Highlighting the intersectionality of movement issues—that is, how gender justice, anti-racism, and environmental justice all overlap—can help more people see how the social issues they care about are connected to the environment.

Understanding that the same systems that oppress the environment also oppress marginalized communities of people—a core insight of ecofeminism—can further contribute to people's capacity to recognize how environmental issues relate to their personal lives. The same logic that elevates culture over nature also elevates men over women, leading to social arrangements where women experience environmental hazards first and worst and, therefore, rise up to propose solutions. This is one reason why women tend to be the majority of activists in environmental movements, particularly at the grassroots level (Bell and Braun 2010; Seager 1996; Stein 2004), a trend that holds in my research.

Though many interviewees had not thought much about how gender informed their organizing, their emphasis on collaboration, care, and community illustrates that feminine perspectives, values, and practices, shaped in particular ways by activists' lived experiences, are important pillars of their activism.[4] These ecofeminist values shape how the grassroots movements that I study understand the problems they face, the solutions they would like to see, and *how* they try to put these solutions into practice. Gender, then, is a structure of inequality, an individual identity, and a force in social interactions that shapes values and ways of relating to other people that are central to how the climate justice movement works within my research cases.

Building strong, broad-based social movements is one method for achieving environmental and climate justice for everyone. Activists and scholars use the term *movement building* to describe the work movements do to inspire people to get involved, feel like they are part of a movement, and collaborate. Movement building includes creating organizations, relationships, networks, skills, identities, frames, and strategies—all the things that are required for sustained mobilization (Juris et al. 2014:329). Movement building is an outcome, just as legal victories or policy changes can be outcomes, of social movements. In other words, social movements can have goals related to movement building, and therefore, may succeed in movement building alongside successes, or failures, in policy or legal change. Movement building is about ensuring everyone has a seat at movement tables to envision, together, a different world.

The movement building that climate justice activists prioritize, and how people actually do movement building, has received less attention than other topics in social movement studies, especially at the grassroots level and in comparative contexts (Blee 2012; Juris et al. 2014). My research enhances understanding of two components of movement building—collective identity and coalition building.

As activists' shared understandings of the context in which they organize and their plans of action, collective identity shapes how "actors 'organize' their behavior, produce meanings and actively establish relationships" (Melucci 1989:36). This conceptualization by Alberto Melucci resonates most closely with my research because of its focus on relationships and because of its understanding of collective identity as a constantly changing process, rather than a static definition. In her analysis of feminist mobilizations in Spain, María Martínez (2018) likewise argues that collective identity is best understood as a complex and unfinished process, grounded in emotions and relationships that people activate and transform repeatedly over time.

Collective identity is particularly important for the environmental and climate justice movements. Shannon Bell (2016) demonstrates that what people perceive as the collective identity of the environmental justice movement has actually deterred people on the front lines of mountaintop-removal coal mining (who, for example, suffer health effects related to coal mining) from participating in environmental justice organizations. They do not identify with those who identify themselves as environmental justice activists. This is troubling for the environmental and climate justice movements because a central tenet of both is that their efforts should be led by the people most affected by climate change or environmental degradation. This is because, in line with ecofeminism, the people experiencing the highest levels of oppression have insights that are critical to designing paths toward justice. In Crenshaw's words, the goal of movements for justice "should be to facilitate the inclusion of marginalized groups for whom it can be said: 'When they enter, we all enter'" (1989:167). Understanding ways to create collective identities that resonate with broad bases of people, especially those on the front lines of climate change and energy extraction, is critical to achieving the climate justice movement's goal of building a movement of *everyone* to change everything.

A second core component of movement building, something upon which a broad-based movement depends, is coalition building. Coalition building is the formation of relationships among people and across organizations. These relationships facilitate people's capacity to draw on material and social resources. They also broaden the scope of issues, perspectives, strategies, and tactics that inform social movements' actions. Appealing to diverse identities and recognizing difference as a strength are key to successful coalitions (Bystydzienski and Schacht 2001; Lipsitz 2006). Previous books on coalitions examine how social networks, ideology, and social and political contexts inform coalition emergence (Van Dyke and McCammon 2010); focus on how particular groups build relationships (see Davis 2010; Grossman 2017 on Native–non-Native alliances); or present edited collections of wide-ranging movements and contexts (e.g. Bystydzienski and Schacht 2001; Van Dyke and McCammon 2010). This book complements these efforts by enhancing understanding of the characteristics of *effective* coalitions, across diverse cases, groups of people, and organizations that are all connected by their common resistance to extreme energy extraction. Rather than asking why coalitions emerge, I explore how they work well, or do not, arguing that effective coalitions are intentionally constructed through the learning, labor, and creativity of activists.

In the tradition of scholarship and community organizing on climate and environmental justice, ecofeminism, and movement building, this book shares knowledge about how people work together to imagine and realize a feminist future and a livable climate. Through attention to each of these concepts, activists can enhance capacity for working across lines and scholars can illuminate processes that contribute to social movements' potential for building inclusive collective identities and coalitions. My hope is that, together, we can build feminist climate justice, a fossil-free and non-capitalist society that does not accept exploitation, where communities, rather than corporations, determine their futures, and where decisions are guided by what the most marginalized communities deem to be in the interest of peace and justice.

METHODOLOGY: POSITIONALITY, ACCOUNTABILITY, AND RECIPROCITY

My interest in working together and resisting fossil fuels is political and scholarly. As a person coming of age during this period, I am concerned about the future. I want to lend my labor to climate justice.

My positionality as a young white woman who spent my formative years in Idaho and California facilitated this research. For example, Alma and her girlfriends (who also resisted natural gas) invited me to their weekend coffee dates where the "girls" would get together to talk about their week. These conversations were mostly about oil and gas. Being in my mid-twenties and a graduate student facilitated my role as a learner. Interviewees, most of whom were significantly older than me, were generally very willing to share their stories. Finally, my own connection to place played an important role in my interactions with research participants. I strongly identify as someone from Idaho. I lived in Idaho from age seven to twenty-one. My connection to Idaho was one reason I wanted to do research there. I wanted to be informed about what was happening and build connections with people trying to preserve a way of life that I hold dear. This connection also facilitated my research, both in terms of previous contacts I had and by making me a kind of insider. Though I was coming from California, I was not "one of *those* Californians" that longtime Idaho residents see as trying to change Idaho.[5] I also look like I am from Idaho, a state with very few people of color. As Bell (2016) has shown, being perceived as an outsider because of where you are from, or because of your opposition to industry and local elites, can be damaging for social relationships. Many of my interviewees played up their "local" or insider status, frequently referencing how long they had lived in Idaho, confirming that "outsider stigma" is real and something to be avoided.

On the other hand, being a California resident, who had built up relationships in the community over years by participating as a core member of the climate justice group 350 Santa Barbara, facilitated my insider status in Santa Barbara County, the site of my research. In this setting, I was not just a graduate student passing through, but someone who had worked alongside grassroots activists for years. They trusted me to tell "our" story. Nonprofit staff also trusted me. They had seen me at local events, county hearings about energy, and knew that I had been involved with Measure P—a countywide effort to ban hydraulic fracturing—from the beginning.

Throughout the research, I identified myself as a scholar activist and strove to practice feminist accountability and reciprocity (see Bhavnani 1993; Haraway 1988; Pulido 1996). I worked to make my writing accessible and useful to activists, asked for interviewee feedback on my writing, and helped out in movement spaces by taking meeting minutes, paying for meals, carrying children, and cleaning. I also practiced accountability by offering interviewees the option of using their real names. Using real names is important because it gives people credit for their ideas and can facilitate movement building by giving readers the ability to learn more about interviewees' work and to perhaps even link up with them and their organizations. Finally, feminist scholar activism means that I hope this research will facilitate social movements, public discussion, and scholarly capacity to envision and build a just and sustainable world that centers the experiences and expertise of communities most marginalized by the status quo.

The data in this book come from ethnographic fieldwork and 106 in-depth interviews. My ethnographic fieldwork consisted of a total of three months conducting research in Idaho in 2015 (spread throughout February, March, July, October, November, and December) and my participation and participant observation in climate justice organizing in Santa Barbara, California, from September 2013 to September 2016.

The interviews were styled as "conversations with a purpose" (Burgess 1984), semi-structured interviews that recognize the importance of establishing relationships of trust and confidence with interviewees. I conducted sixty-two interviews in Idaho during the months I conducted fieldwork in 2015. In Santa Barbara, I conducted a total of forty-four interviews from May 2015 through September 2016. Twenty-nine of these interviews were with youth activists. Interviewees' ages ranged from nineteen to seventy-eight years old. They identified as white, Native American, Latina/o, African American, Black, Filipino, Sinhalese Sri Lankan, and as mixed race and biracial. They held positions in different types of organizations. Sixty-two percent of all interviewees were women, reflecting their overwhelming participation and leadership in the organizations I engaged with. Interviews averaged seventy-two minutes in length and explored themes including interviewees' work as activists, their perspectives on diversity and inclusivity in their groups, and their hopes for the future.

RESEARCH CONTEXT

My research is set within a contradiction. In the late 2010s, climate science demanded we keep fossil fuels in the ground while US politics continued to support the extraction of the dirtiest fossil fuels. Global and national concern about the climate crisis resonated with the science. In 2016, the Paris Agreement went into effect, spurring the global community toward coordinated efforts to reduce greenhouse gas emissions. That year, 64 percent of Americans worried a fair amount or great deal about global warming, the highest percentage since 2008 (Saad and Jones 2016). In 2018, 70 percent of Americans believed global warming was happening, with 85 percent in support of funding renewable energy research (Marlon et al. 2018). During this same period, the United States ignored these trends, becoming the world's leading producer of oil and gas and ramping up its use of extreme extraction techniques.[6] These include fracking, which uses many more resources and produces more greenhouse gas emissions than conventional drilling methods (US Energy Information Administration 2016b; 2016c; 2016d). Concerned communities rose up in resistance across the country, not only against extraction, but also against the transportation of extreme energy like the Canadian tar sands.

Amid this broader context, this research is grounded in California and Idaho, two places with different political, energy, and social contexts. In California—historically the most important and progressive state for environmental and climate justice—my research centers on Santa Barbara County, the site of the world's first offshore oil drilling in 1896, the United States' first major oil spill in 1969, and the birth of the modern American environmental movement. In 2014, amid a historic drought, California's warmest year ever at the time, and proposed oil expansion, grassroots climate activists in the group 350 Santa Barbara formed a coalition that attempted to ban extreme energy extraction in the county with a ballot measure, Measure P. The effort relied upon substantial organizing by youth activists at Santa Barbara's college and university. Over six million dollars in oil industry opposition inundated the community, sharpening already existing divisions among residents, especially political and racial divides. Despite the measure's failure on election day, the struggle remains the largest county-level electoral mobilization (in terms of volunteers and money) to ban fracking in California history.

Idaho is a more politically conservative state, with no oil or gas extraction until 2009. While often ignored, decisions in Idaho and similar states in the center of the country impact the majority of US land and, when combined, shape federal policy. My research in Idaho is based in three regions: southwest, central, and northern Idaho. Southwest Idaho is the site of the state's nascent natural gas industry. In this region of rural farms surrounding Idaho's capital and largest city, residents formed the group Citizens Allied for Integrity and Accountability (CAIA) in 2015, which is engaged in an ongoing struggle to protect property rights. Representatives of Idaho's small number of statewide environmental nonprofit groups, after attempting to strengthen Idaho's natural gas regulatory structure, mostly stand on the sidelines, unsure of how to widen their mission statements to include concerns of property-rights activists who do not consider themselves to be environmentalists. In the other two regions included in my research, central and northern Idaho, from 2011 to 2014, a grassroots coalition of individuals and organizations protested two-hundred-foot-long megaloads (trucks with trailers) of tar sands infrastructure on rural highways and in small towns before successfully barring the loads in a legal suit that concluded in 2017.

In both states, activists were in conversation across campaigns and across geographies, with many participating in regional, state-wide, national, and global social movement communities. My analysis elucidates the character of resistance to extreme energy extraction in these two different states to highlight commonalities across diverse contexts and to show how the particularities of these contexts inform how people attempt and fail to work together.

ORGANIZATION OF THE BOOK

Chapter 1 lays the foundation for the remainder of the book by detailing the politics of climate change, extreme energy extraction, and the climate justice movement. I juxtapose the crisis of climate change and the need to keep fossil fuels in the ground with the continued pursuit of fossil fuel energy, particularly in the United States. In this context, the extraction, transportation, and processing of extreme forms of energy, such as tar sands and hydraulically fractured oil and gas, have taken off, exacting tremendous environmental and human costs. I present climate justice and the climate justice movement as one way to confront this situation, providing the reader with background on the movement and its goals.

Chapter 2 provides the reader with further information on the two research contexts of Idaho and California and the specific towns where interviewees and their campaigns are based. I provide a concise overview of how the two states have managed fossil fuels and the key actors and campaigns that are the focus of the book.

Tracing interviewees' journeys into activism, chapter 3 provides an in-depth account of how rural southwestern Idahoans built an unlikely alliance to resist natural gas development. The group they formed, called Citizens Allied for Integrity and Accountability (CAIA), has had board members who were former leaders of the local Tea Party group, people who consider themselves Democrats, climate change skeptics, and people who are very concerned about climate change. I argue that this unlikely trans-partisan alliance is created and maintained by the practice of *talking across lines*. Rather than rally around what have become divisive political issues, such as climate change, CAIA focuses on issues of private property, public infrastructure, and, as its name suggests, government and corporate accountability and integrity. The activism journeys I share in this chapter help readers understand activists' lived experiences and values, the social setting in their communities, and why activists have needed to develop the tactic of talking across lines.

Chapter 4 analyzes how people succeed and fail in their efforts to talk across political lines in the context of the fight against natural gas in Idaho. Talking across lines depends on building relationships of trust through a dedication to shared values—in mission statement, messaging, and engagement—within and beyond the group. It also requires acknowledgment of differences in perspective by complicating labels related to political party, climate change beliefs, activism and Not In My Back Yard (NIMBY) environmentalism, and assumptions about how and whether these labels inform a person's willingness to resist natural gas. Ultimately, talking across lines requires that activists agree to disagree on certain issues in the interest of working together to advance shared goals, a complicated way of navigating difference that is heavily reliant on trust.

Chapter 5 focuses on Santa Barbara, California, to explore the values and practices of youth climate justice activists. These youth, primarily organizing in the context of groups and campaigns embedded in college and university environments, are developing progressive values and practices that create a particular culture of organizing. I call this culture a *climate justice culture of creation* because it is a political culture focused not only on resistance, but also on creation (see Foran 2014). It prefigures the world that youth want to see in the present—in their groups, interpersonal interactions, and experiences. The core values of this culture are accessibility, intersectionality, relationships, and community. Youth strive to make organizing accessible and enjoyable to all. They teach each other how social and environmental inequalities intersect to inform people's lived experiences and different passions. Prioritizing relationships as the basis of understanding and supporting each other, they envision healthy communities as those where people are politically engaged and willing to build relationships with people who do not share their views. Specific practices, including horizontal leadership structures, anti-oppression trainings, and work to diversify members and leaders all embody these values. I see youth's climate justice culture of creation as a powerful movement-building tool for working across lines.

In chapter 6, I transition to comparative analysis of my cases across campaigns, locations, and activist groups. The chapter explores the tensions and possibilities for building coalitions between grassroots and staffed nonprofit organizations—what I call grasstops organizations. I find that grasstops' commitment to pragmatism challenges activists' efforts to build coalitions between these elements of the movement. This pragmatism in rooted in grasstops' organizational form, particularly nonprofits' responsibilities to fulfill mission statements and secure funding, as well as strategic, tactical, and motivational divergences. To bridge these divides, activists stress the importance of welcoming new ideas, giving attribution, and centering, rather than marginalizing, demands for radical systems-changing actions.

Chapter 7 explores two resistance efforts—the Measure P effort to ban fracking in Santa Barbara County, and the mobilization to stop the transportation of tar sands processing equipment on giant trucks or "megaloads" on Idaho highways. I narrate each story of struggle, providing a synthesis of interviewee perspectives and lessons learned. I discuss the diversity of concerns and tactics that made the megaload struggle successful and the strengths and weaknesses that characterized the impressive and yet ultimately unsuccessful Measure P electoral campaign.

Chapter 8 employs comparative analysis to present the factors that facilitated and inhibited working across lines in the Measure P and megaload campaigns—factors that shaped the different outcomes of the campaigns. It considers how the differing presence and power of oil, support for environmental politics, connection to place, and campaign durations informed these resistance efforts. Despite these differences, there were also many similarities rooted in shared perceptions of urgency and values. Activists in both places offered ideas for best practices for working across lines that resonate with the other cases in this book: horizontal and relational organizing to build trust and meaningful participation of all involved. The final section of this chapter examines commitment to diversity and inclusivity, of organizing tactics, and of participants of many ages and of diverse racial, ethnic, and Indigenous backgrounds. It draws lessons from the successful collaboration of Nez Perce and non-Native people in Idaho, and Measure P's inability to build bridges between white environmentalists and the Santa Barbara County Latinx community.

The conclusion reviews the analysis and argument, illuminates the contributions of this research, and suggests paths forward. Closing with interviewees' hopes for the future, I invite the scholarly community to build knowledge on how working across lines happens in other contexts, and the general reader to build relationships, cultivate values, and share practices that facilitate social transformation toward a fossil free, community-centered world.

I hope this book nourishes your desire, capacity, and joy for working across lines to meet the challenges we face.

CHAPTER I

The Energy and Political Landscape

Climate Crisis, Extreme Energy, and the Climate Justice Movement

Risks are considered key due to high hazard or high vulnerability of societies and systems exposed, or both. [. . .] The key risks that follow, all of which are identified with *high confidence*, span sectors and regions. [. . .]

1. Risk of death, injury, ill-health, or disrupted livelihoods in low-lying coastal zones and small island developing states and other small islands, due to storm surges, coastal flooding, and sea level rise.
2. Risk of severe ill-health and disrupted livelihoods for large urban populations due to inland flooding in some regions.
3. Systemic risks due to extreme weather events leading to breakdown of infrastructure networks and critical services such as electricity, water supply, and health and emergency services.
4. Risk of mortality and morbidity during periods of extreme heat, particularly for vulnerable urban populations and those working outdoors in urban or rural areas.
5. Risk of food insecurity and the breakdown of food systems linked to warming, drought, flooding, and precipitation variability and extremes, particularly for poorer populations in urban and rural settings.
6. Risk of loss of rural livelihoods and income due to insufficient access to drinking and irrigation water and reduced agricultural productivity, particularly for farmers and pastoralists with minimal capital in semi-arid regions.

7. Risk of loss of marine and coastal ecosystems, biodiversity, and the ecosystem goods, functions, and
   services they provide for coastal livelihoods, especially
   for fishing communities in the tropics and the Arctic.
8. Risk of loss of terrestrial and inland water ecosystems,
   biodiversity, and the ecosystem goods, functions, and
   services they provide for livelihoods.

Many key risks constitute particular challenges for the least
developed countries and vulnerable communities, given their
limited ability to cope.

—Climate Change 2014: Impacts, Adaptation, and Vulnerability.
Summary for Policymakers. Working Group II Contribution to the
Fifth Assessment Report of the Intergovernmental Panel on Climate
Change

President Trump is committed to eliminating harmful and
unnecessary policies such as the Climate Action Plan [. . .
and] will embrace the shale oil and gas revolution.

—An America First Energy Plan, The White House, 2017

Why we're marching. [. . .] Donald Trump's election is a
threat to the future of our planet, the safety of our communities, and the health of our families. [. . .] If the policies he
proposed on the campaign trail are implemented, they will
destroy our climate, decimate our jobs and livelihoods, and
undermine the civil rights and liberties won in many hard
fought battles. It's up to us to stop that from happening
before it starts. [. . .] Join us on April 29th.

—People's Climate Movement, 2017

In the midst of climate crisis, there is a disconnect between science and
policy. The Intergovernmental Panel on Climate Change (IPCC) rec-
ommends a 45 percent reduction in greenhouse gas emissions by 2030
(IPCC 2018). Rather than heed the calls of climate scientists, governments and corporations have been doubling down to extract fossil

fuels—scraping the bottom of the barrel for oil and gas in its dirtiest, most hard-to-reach forms. Since the early 2000s, the climate justice movement has been working to fill the void between the known risks of climate change, which are only becoming more likely, and inaction. The movement has been making the links between the health of the planet and people, as the People's Climate Movement's call to action illustrates. While the contrast between the observations of the Intergovernmental Panel on Climate Change and the America First agenda of the Trump administration is extreme, it is not uncharacteristic of policy and action at local and global levels around the world since 1992, when countries first joined the United Nations Framework Convention on Climate Change (UNFCCC).

This chapter lays the foundation for the remainder of the book by detailing the politics of climate change, extreme energy extraction, and the climate justice movement. The chapters that follow build upon these insights to not only deepen the perspectives of scholars, activists, and the public about this movement, but also to prepare readers with concrete cases and practices that they might adapt to their own efforts to build sustainable communities.

CLIMATE CRISIS

As the opening quote from the Intergovernmental Panel on Climate Change's Fifth Assessment Report communicates, we, humans and nonhumans, are in climate crisis. Despite the reserved scientific language of risks and probabilities in the quote, the emergency of our situation is not far from the surface. "High confidence" is strong language for climate scientists. Key climate risks—mortality, loss of livelihood, increased illness, and loss of ecosystems—become more and more likely as greenhouse gas emissions and temperatures continue to rise (IPCC 2014). These risks will affect people everywhere, especially the most marginalized communities, and the scale of action to mitigate these risks does not match their severity.[1]

From the late 1800s to 2020, humans raised the global average temperature of the earth by more than two degrees Fahrenheit (1.2 degrees Celsius) (NASA 2021). Temperatures during 2011–2020 were warmer than those of any period in the last one hundred twenty-five thousand years (IPCC 2021). And 2020 tied 2016 as the hottest year ever recorded (since 1880). While the twenty-first Conference of the Parties (COP) to the UNFCCC's adoption of the Paris Agreement in 2015 made some strides toward lowering emissions, it did not do enough. The International Energy Agency (2015) reports that implementing the climate pledges outlined in the Paris Agreement would limit the rise in global average temperature by 2100 to 2.7 degrees Celsius. These pledges are voluntary and lack enforcement mechanisms. The last time the world had temperatures of three degrees Celsius above preindustrial levels was three million years ago. Gavin Schmidt, director of NASA's Goddard Institute for Space Studies explained that, "At that time, there was almost no ice anywhere. The sea level was twenty meters (sixty-five feet) or so higher" (Lewis 2015). Between 25 and 92 percent of the population of fourteen of the world's megacities would be under water with a six-meter (twenty-foot) rise, which scientists predict at only two degrees Celsius of warming (Strauss 2015). In the International Energy Agency's main climate scenario, the entire carbon budget for a two-degrees-Celsius future is used up by the early 2040s. The carbon budget is the amount of carbon that can be burned while staying under the globally agreed upon threshold for maximum warming, two degrees Celsius (3.6 degrees Fahrenheit). Bill McKibben (2012), cofounder of 350.org, has argued that simple math, according to this budget, means that fossil fuel corporations hold over five times the carbon budget in their reserves. Therefore, staying within the carbon budget requires that fossil fuel companies keep most of their reserves in the ground.

Keeping fossil fuels in the ground does not sit well with the profit-seeking agenda of the fossil fuel industry. In the decades following World War II, there were seven major multinational companies dominating the oil industry. Known as the seven sisters, they were Royal Dutch Shell, British Petroleum, Gulf Oil, Exxon, Mobil, Texaco, and Chevron (Sampson 1975). In 1960, Saudi Arabia, Iraq, Iran, Kuwait, and Venezuela formed the Organization of Petroleum Exporting Countries (OPEC). Now composed of thirteen nations, OPEC is an intergovernmental organization that unifies the petroleum policies of its member countries, which "exercise permanent sovereignty over their natural resources in the interest of their national development" (OPEC 2017). In the realm of publicly traded companies, the seven sisters, some of which have merged, are still powerful and well known, but they are now joined by major companies from other countries. In terms of revenues, profits, assets, and market value, US-based ExxonMobil was the world's largest publicly traded oil company in 2016, followed by China's state-controlled PetroChina, Chevron (United States), Total (France), Sinopec (China), and Royal Dutch Shell (Netherlands) (*Forbes* 2016a). In terms of 2016 production, Russia's Gazprom and Rosneft led, with Exxon-Mobil in third, followed by PetroChina, BP (United Kingdom), Royal Dutch Shell, Chevron, and Petrobras (Brazil) (*Forbes* 2016b).

The fossil fuel industry's mission is to increase profits for share-holders by finding and selling as much fossil fuel as possible. Internal documents dating from the early 1990s to the mid-2000s demonstrate that this mission led Exxon (now ExxonMobil) and other fossil fuel companies to spend millions of dollars to spread doubt about climate science (InsideClimate News 2015; Oreskes and Conway 2010; Union of Concerned Scientists 2015). They followed the lead of the tobacco industry, which obfuscated the health effects of smoking for decades (Oreskes and Conway 2010; Union of Concerned Scientists 2007). For example, an internal Exxon memo titled "The Greenhouse Effect," from August 1988, noted the scientific consensus on the role fossil fuels play in climate change, concluding: "Exxon Position: Emphasize the uncertainty in scientific conclusions regarding the potential enhanced greenhouse effect" (Jennings, Grandoni, and Rust 2015). The fossil fuel industry has also funded scientists and think tanks to spread doubt about climate science. In one period during 2001 to 2012, fossil fuel interests including ExxonMobil, the American Petroleum Institute, the Charles Koch Foundation, and Southern Company, a utility company that generates most of its power from coal, paid Wie-Hock Soon, a purportedly independent contrarian climate scientist of the Harvard-Smithsonian Center for Astrophysics, more than $1.2 million (Union of Concerned Scientists 2015). He failed to disclose this conflict of interest in most of his scientific papers (Gillis and Schwartz 2015). In another example, the Heartland Institute, which hosts conferences and publishes reports denying climate change (see Klein 2014), received $551,500 from ExxonMobil from 1998 to 2005, 40 percent of which was desig-nated for climate change projects (Union of Concerned Scientists 2007). Dunlap and Jacques (2013) find that 72 percent of climate denial books published from 1980 to 2010 were linked to conservative think tanks (the authors consider the Heartland Institute, Competitive Enterprise Institute, the CATO Institute, and the Marshall Institute some of the leading conservative think tanks behind climate denial).

The fossil fuel industry funded climate-change-denying scientists despite its own sophisticated grasp of climate science since the 1970s (Jerving et al. 2015; Oreskes and Conway 2010; Union of Concerned Sci-entists 2015). While the entire American Petroleum Institute, with mem-bers from nearly every major US and multinational oil and gas company including Exxon, Mobil, Amoco, Phillips, Texaco, Shell, Sunoco, Sohio, and Chevron's predecessors Standard Oil of California and Gulf Oil, was involved, Exxon led the way (Banerjee 2015). Exxon had the largest and most ambitious climate research program, focused on climate modeling, and was the first fossil fuel company to launch campaigns to cast doubt on climate science and stall greenhouse gas regulations (Banerjee 2015). In recent years, the fossil fuel industry has continued this practice, funding think tanks and scientists, as mentioned, as well as political campaigns. In California alone, the fossil fuel lobby (composed of a number of companies including Chevron, Phillips 66, Tesoro, ExxonMobil, and AERA Energy) spent over $32 million in the 2015–2016 legislative session (American Lung Association in California 2016). Alongside its political wrangling to obscure climate science, the industry has been ramping up extraction of especially hard-to-reach fossil fuels through unconventional techniques.

EXTREME ENERGY EXTRACTION

Extreme energy extraction techniques have spread throughout the world. They are used to extract coal, oil, and gas—all major forms of fossil fuel. In Appalachia in the last few decades, the coal industry has been removing mountain tops, literally, to expose coal seams. In the process of mountaintop-removal (MTR), explosives equivalent to the power of one Hiroshima bomb are detonated each week (Cho 2011). Twenty-story machines called draglines, which can move seven dump trucks worth of soil per scoop, then remove the mountain, dumping the debris into surrounding valleys (Perks 2009). More than 502 peaks have been leveled—an area about the size of Delaware (Perks 2009). Two thousand miles of Appalachian headwater streams have been buried (US EPA 2011). While federal law requires that these sites be restored following extraction, such a feat is impossible—habitats created over millions of years cannot be "put back." Companies rarely attempt to return the areas to their original surface configuration because they often receive waivers from state agencies with the idea that economic development will occur on the newly flattened land (Appalachian Voices 2013). Yet 89 percent of MTR mines are not used for economic development beyond forestry or pasture; most previous mine sites remain undeveloped (Geredien 2009).

The scale of extraction of the Alberta tar sands is even larger than MTR. Former director of NASA's Goddard Institute for Space Studies

James Hansen has said that if humans were to extract and burn the carbon in the tar sands, it would be "game over for the planet" (Hansen 2012). The Canadian nonprofit organization Environmental Defence deems it "the most destructive project on earth" (Hatch and Price 2008) and Black et al. (2014b) offer a number of websites where, reminiscent of J. R. R. Tolkien's *Lord of the Rings* trilogy, a viewer can visualize this "Mordor Landscape" (8). Similar to MTR, tar sands pilot operations in Alberta have been in progress since the 1960s, with production vastly expanding since the 1990s when the specter of "peak oil" made oil prices sufficiently high to justify this expensive and resource-intensive process (Black et al. 2014b).

The tar or oil sands are a crude heavy oil substance called bitumen. This tar-like substance is mixed with sand, clay, and water, making it very hard to get out of the ground. The Athabasca River Basin in western Canada contains the world's largest deposit of the substance, which is refined into products like gasoline after extraction. It can be extracted through surface mining—where the bitumen is excavated—and through the energy intensive process of "in-situ" mining.[2] Used for extracting deep deposits of bitumen, which comprise 80 percent of reserves, in-situ mining uses steam, injected under high pressure, to liquefy the bitumen so that it can be extracted (Union of Concerned Scientists 2013). Waste water from the process—quantities of three to five cubic meters for every cubic meter of extracted bitumen—are stored in pits called tailings ponds that are so large they can be seen from outer space (Black et al. 2014b:9).[3] The 2013 annual carbon emissions from the extraction and burning of bitumen in Alberta were estimated to be more than the combined emissions of one hundred nations (Saxifrage 2013).[4] Added to this are methane emissions from tailings ponds and carbon emissions released from the destruction of peatland (Black et al. 2014b:15). Through fossil fuel energy used for mining and refining, and the landscape changes that go along with these processes, tar sands extraction has tremendous impacts on local environments and the climate.

Hydraulic fracturing, or "fracking," is the most recently contested form of extreme extraction. Since 2000, also in response to higher oil and gas prices and technological advancements (Brown and Yucel 2013), fracking has boomed across the United States (US Energy Information Administration 2016c; 2016d). In 2000, fracking produced 3.6 billion cubic feet per day (Bcf/d) of marketed gas, accounting for just 7 percent of total US natural gas production; in 2015, fracking produced more than 53 Bcf/d, accounting for 67 percent of total production (US Energy

Information Administration 2016c). The percentage of oil produced through fracking in the United States increased from 2 percent in 2000 (equivalent to 102,200 barrels per day (b/d) of oil) to about 50 percent (equivalent to 4.3 million b/d) in 2015 (US Energy Information Administration 2016d). Fracking is occurring in twenty-one US states and is expanding (Horn 2016). Exact well counts are difficult to assess and can differ by source as there are no national standards for publishing oil and gas data (Kelso 2015; US EPA 2016).

The most prolific regions for fracking are the Bakken (covering portions of North Dakota and Montana), Eagle Ford (Texas), Haynesville (mostly in Texas and Louisiana), Marcellus (New York, Pennsylvania, and West Virginia), Niobrara (mostly in Colorado and Wyoming), Permian (Texas and New Mexico), and Utica (Ohio) formations (US Energy Information Administration 2017).[5] These regions accounted for 92 percent of oil production growth and all natural gas production growth in the United States from 2011 to 2014 (US Energy Information Administration 2017). For example, between 2005 and 2016, North Dakota, which sits atop the Bakken formation, increased its oil production tenfold; 70 percent of this increase occurred from 2011 to 2014 (US Energy Information Administration 2016b). Since 2017, the Anadarko formation in Oklahoma has joined the list of major producers (US Energy Information Administration 2021a). Fracking is also expanding globally. Radetzki and Auilera (2016) estimate that the United States has only 17 percent of the global shale oil share. Fracking can be used to extract oil and natural gas from shale (a type of sedimentary rock) and other tight rock formations, as well as to improve yield from conventional oil and gas fields (Radetzki and Auilera 2016).

The process of fracking, which can occur in vertical or horizontal wells that can extend over a mile down and a mile horizontally from the well head (FracTracker Alliance 2017), injects large quantities of water, sand, and a chemical slurry to break up the rock formation and allow gas or oil to escape. A congressional report on the chemicals used in fracking by the fourteen leading oil and gas service companies found that the companies used twenty-nine chemicals that were known or possible human carcinogens, regulated under the Safe Drinking Water Act for their risks to human health, or listed as hazardous air pollutants under the Clean Air Act (Committee on Energy and Commerce 2011). These included methanol, ethylene glycol, diesel, xylene, hydrogen chloride/hydrochloric acid, toluene, ethylbenzene, formaldehyde, and sulfuric acid (Committee on Energy and Commerce 2011). During fracking, these chemicals mix with water, producing wastewater. In California, where fracking extracts heavy tar-like oil, each hydraulically fractured well produces ten or more gallons of wastewater for every gallon of oil produced (Cart 2015c). This water is then injected into the ground, often into aquifers with water classified as clean for human consumption (Cart 2015c). In 2016, the United States Environmental Protection Agency released the first nationwide study on the impacts from fracking on US drinking water resources, identifying cases of negative water impacts throughout all stages of the hydraulic fracturing water cycle. It is worth noting that more studies on fracking were published in 2014 than from 2009 to 2012 combined (Concerned Health Professionals of NY and Physicians for Social Responsibility 2016). Infant deaths and birth defects are just two of the emerging health outcomes correlated with drilling and fracking operations (Concerned Health Professionals of NY and Physicians for Social Responsibility 2016).

By enabling a boom in oil and natural gas production, fracking also exacerbates climate change. Natural gas is often hailed as a "bridge fuel" to renewable energy, something that is "cleaner," in terms of greenhouse gas emissions, than coal (Plumer 2014). In terms of carbon dioxide, it is cleaner, emitting half as much carbon as coal when burned. Focusing on carbon dioxide, however, misses the whole picture (McKibben 2016a). Natural gas is mostly composed of methane. When not burned, methane traps heat in the atmosphere much more efficiently than carbon dioxide. Unlike carbon dioxide, which lasts centuries, methane lasts just one or two decades. Over one or two decades, however, it is between 86 and 105 times more potent as a heat-trapping greenhouse gas than carbon dioxide (McKibben 2016a). US methane emissions increased by more than 30 percent over the 2002 to 2014 period (Turner et al. 2016), largely as a result of the simultaneous boom in the fracking industry (McKibben 2016a). For as Howarth, Santoro, and Ingraffea (2011) find, between 3.6 and 7.9 percent of methane gas from US shale drilling operations escapes into the atmosphere. These authors conclude that shale gas has a greenhouse gas footprint greater than all other fossil fuels—a footprint at least twenty percent greater than coal and perhaps more than twice as great on the twenty-year horizon (Howarth, Santoro, and Ingraffea 2011).

Extreme energy extraction not only contaminates the ground, water, air, and climate, but it is also linked to social degradation. MTR results in unstable mountains that can easily slide over homes during heavy rains, and, because of increased mechanization and de-unionization,

MTR also decreases employment opportunities and social capital (Bell 2013; 2016; Scott 2010). Fracking increases truck traffic, decreases revenues and enjoyment from tourism and recreation because of a destroyed landscape, and creates a boom-and-bust economy associated with higher crime rates, substance abuse, sexual assault, mental illness, inadequate housing, and overextended public services (Food and Water Watch 2013:4). Farmers face corporate bullying as industry relies on procedural inequities, related to lease negotiation and enforcement, to expand extraction (Malin and DeMaster 2016). Relative to unfracked counties, fracked counties in Pennsylvania—atop the Marcellus Shale—have had substantial increases in truck crashes (some of which spill frack water into surface water), disorderly conduct arrests, and cases of sexually transmitted diseases (Food and Water Watch 2013). Social disorder crimes increase because socially isolated oil and gas workers have "ample income and little to occupy their time in rural communities," with many turning to alcohol (Food and Water Watch 2013:7). In parallel to Pennsylvania, North Dakota sees increased rates of domestic violence, and local women report feeling unsafe as a result of fracking "man camps" filled with young men from other states; as a result, North Dakota has one of the highest male-to-female ratios in the United States (Eligon 2013). On the Fort Berthold Reservation in North Dakota, Native women have experienced exponentially increasing rates of violence linked to fracking man camps (Honor the Earth n.d.), heightening the crisis of missing and murdered Indigenous women and relatives (see Urban Indian Health Institute 2019).

Gender-based violence related to man camps exists in the tar sands region as well (Awâsis 2014; Thomas-Muller 2014), where the land and animals that shape First Nations' cultural identities have been decimated. Animals that First Nations depend on for food have heightened levels of environmental contaminants, including arsenic, cadmium, mercury, selenium, and polycyclic aromatic hydrocarbons (PAHs) (McLachlan 2014). As Lucas (2004) demonstrates, toxins that build up in animals become even more concentrated in humans who eat the animals. Some First Nations people have stopped eating their traditional foods and drinking from traditional water sources for fear of contamination (Lameman 2014; McLachlan 2014). They then must rely on store-bought foods that offer lower nutritional value. Their fears are well founded. McLachlan (2014) found that twenty of ninety-four participants in a health study downstream of the tar sands had experienced cancer. He concluded that cancer occurrence increased with consumption of traditional food and locally caught fish and with employment in the tar sands.

The extraction of tar sands goes against treaty rights of First Nations communities (Awâsis 2014; Lameman 2014). For example, the Athabasca Chipewyan First Nation have charged that Shell Oil's mining of the tar sands violates the 1899 Treaty 8, between First Nations in northern Alberta and Queen Victoria, which protected their right to practice traditional lifeways (Moe 2012). The treaties violated by the Enbridge Line 9 pipeline to transport tar sands, under challenge by First Nations (Bueckert 2016), include the Nanfan Treaty, Kaswentha, the Great Peace of Montreal Treaty, the Royal Proclamation of 1763, the 24 Nations Treaty of 1764, and the Haldimand Proclamation (Awâsis 2014). By challenging treaty violations, First Nations can pursue some of the potentially most powerful legal avenues to stop tar sands projects (see Black et al. 2014a).

Transportation of extreme energy has also spurred resistance. The Keystone XL Pipeline was an important catalyst for the climate justice movement in the United States (Russell et al. 2014). Slated to transport tar sands from Alberta, Canada, through the center of the United States to refineries on the Gulf Coast, it met opposition from a diverse array of Native communities, landowners, and environmentalists along its proposed route. Over two weeks in 2011, 1,250 people were arrested in front of the White House protesting the pipeline. This demonstration joined countless local protests that occurred across the country until, in November 2015, President Obama rejected the northern leg of the pipeline. This was a major victory for many members of the movement, despite the fact that in 2012, Obama had fast-tracked the southern leg of the pipeline through Texas and Oklahoma (see Foytlin et al. 2014).

In the midst of the Keystone XL struggle, the oil industry turned to railways to get its product to global markets. In doing so, it created what activists call "bomb trains." A national phenomenon, the transportation of oil by rail grew 4,200 percent between 2008 and 2013 (Association of American Railroads 2014). Matching this trend were increases in explosions. In Quebec in 2013, a wreck decimated the small town of Lac-Mégantic and killed forty-seven people. Major accidents have occurred in Oklahoma (2008), North Dakota (2013), Alabama (2013), Virginia (2014), West Virginia (2015), and Ontario (2015). Rail accidents spilled 800,000 gallons of crude oil from 1975 to 2012, and, in 2013 alone, this same type of accident led to spillage of 1.15 million gallons of crude oil (Warner and Kaine 2014). While opponents of extreme energy have a chance to stop construction of new pipelines that they fear will leak, train tracks are much more plentiful and already serving as pathways for the transportation of dangerous materials through the hearts of communities.

In 2016, a second large-scale pipeline resistance movement, centered on asserting Indigenous sovereignty, rose to meet the Dakota Access Pipeline, built to transport fracked oil from the Bakken Shale Formation in North Dakota to refineries in Illinois. Originally slated to run near Bismarck, North Dakota, the pipeline was rerouted to cross just north (within a mile) of the Standing Rock Sioux Reservation and through unceded Sioux territory, lands that are legally theirs, according to treaties that the US government has repeatedly violated (see Whyte 2017). The reroute occurred despite the Standing Rock Sioux Tribe's opposition to pipelines since 2012. The two counties that comprise the Standing Rock reservation have some of the highest poverty rates in the country (US Census Bureau 2019b). The county where Bismarck is located has one of the lowest poverty rates and is majority white. The US Army Corps of Engineers rejected the Bismarck route because of the pipeline's potential impact on municipal water supplies (Dalrymple 2016). In a blatant violation of treaty rights and sovereignty, the pipeline instead went underneath Lake Oahe, the drinking water source of the Standing Rock Sioux Tribe, which is within their territory. Against this threat, tribal members calling themselves water protectors invited people to their land to assert their rights and defend the water and ancestral sites in summer 2016.[6] Native youth and women began and led the resistance.

In the following months, thousands of people flocked to Standing Rock, where they organized living quarters, workshops, distribution of aid, and direct action. They braved dogs, water cannons in freezing temperatures, and the onset of the winter in North Dakota (see Democracy Now! 2016; Goodman 2016). On December 4, 2016, after thousands of veterans had arrived at Standing Rock to join the water protectors, the Army Corps of Engineers denied an easement to Energy Transfer Partners, the pipeline company, and ordered an environmental impact statement and exploration of alternate routes for the pipeline. On the heels of this victory, the camp quickly shrunk. Just over a month later, however, on January 24, 2017, newly elected President Trump signed an executive order restarting the Dakota Access and Keystone XL pipelines. Armed police evicted the remaining water protectors and destroyed their camp at Standing Rock in February 2017. Oil began flowing in the Dakota Access Pipeline (DAPL) four months later. While President Biden responded to movement demands to stop Keystone XL with a 2021 executive order that prompted the pipeline company TC Energy to cancel the project, Biden's Army Corps of Engineers has opposed shutting down the Dakota Access Pipeline. Therefore, the future of DAPL remains uncertain and will likely play out in the courts.

These cases of direct-action resistance shape the context in which my interviewees live and organize. Many of my interviewees protested the Keystone XL Pipeline before I interviewed them, engaged in resistance to fracking and oil by rail while I conducted research, and, after I finished conducting interviews, mobilized against the Dakota Access Pipeline. These are the campaigns through which activists are developing values and practices, experimenting, connecting with each other, and learning as they work to address climate crisis and create climate justice.

CLIMATE JUSTICE: A SOCIAL RESPONSE TO CAPITALISM'S CONTRADICTION

Climate crisis, and the extreme energy extraction that exacerbates it—perhaps alongside nuclear weapons—pose the greatest large-scale threat to civilization in recorded history. The juxtaposition of extreme energy extraction and climate crisis also clarifies a contradiction inherent to capitalism. Endless growth is not compatible with a livable physical environment, just as extreme energy extraction is not compatible with the implications of climate science—that greenhouse gas emissions must decrease. Climate crisis, then, is a symptom of capitalism in crisis (Clark and York 2005; Foster, Clark, and York 2010; Klein 2014). As the fossil fuel industry continues its work to increase profits through geographical expansion (both on and below the surface of the earth), it decreases the capacity of the earth to sustain capital or labor. A stark example is Arctic oil drilling, which would emit more climate-changing greenhouse gases in the Arctic, now accessible because of melting sea ice (see Rosenthal 2012).[7] As the fossil fuel industry expands exploitation of fossil fuels in the Arctic and elsewhere, it exacerbates the climate crisis, which will eventually endanger the very consumers upon which fossil fuel companies (and capitalism) depend. Understanding that solutions to climate crisis that are socially just and ecologically sustainable require something other than capitalism is a core principle of climate justice.

Climate Justice

Karl Marx wrote that "to be radical is to grasp things by the root" ([1844] 1978:60). Climate justice identifies the root causes of climate change as capitalist social relations, relations of inequality that exploit people and the environment to accumulate profit for a few.

The concept of climate justice was created by coalitions of individuals and organizations at global meetings.[8] Intimately sociological in its clarification of the interconnections between different structures of inequality, climate justice recognizes that social justice requires a livable world, and a livable world can only emerge through social justice. Climate justice centers anti-racist environmentalism, system transformation, and the notion of ecological debt—that the Global North owes a debt to the Global South for its disproportionate contributions to climate change (Bond 2014).[9] Climate justice seeks to address not only current injustices, but also those of the past. Enshrined in the phrase "common but differentiated responsibilities" that is central to the United Nations Framework Convention on Climate Change (United Nations 1992), past injustices refer to the Global North's disproportionate use of the carbon budget—the amount of carbon that can be released into the atmosphere while staying beneath the United Nations' threshold of two degrees Celsius (McKibben 2012)—to industrialize. For example, the United States and European Union emitted 52 percent of the world's total carbon dioxide emissions from 1850 to 2011 (World Resources Institute 2014).

Extreme energy extraction is a clear environmental and climate justice problem. Corporations that perpetuate extreme energy extraction remain profitable by effectively drafting their own regulations that externalize the consequences and costs of putting chemicals and greenhouse gases into the environment, especially upon marginalized communities. For example, in 2005, Halliburton successfully lobbied for fracking to be exempt from the Safe Drinking Water Act of 1974. Fracking operations are made possible through the use of carcinogenic chemicals that corporations do not have to disclose to regulators or the public because they are considered trade secrets (Committee on Energy and Commerce 2011). As extreme energy procedures contaminate water supplies (Cart 2015c) and negatively affect the health of local residents, especially children (McKenzie et al. 2014), they simultaneously exacerbate climate change through their greenhouse gas emission-intensive operations and their perpetuation of an energy system dominated by fossil fuels.

The interconnections of injustice, extreme energy extraction, and climate change have been spreading geographically. The fossil fuel industry's search for more reserves has multiplied the number of places where fossil fuel extraction is either planned, or beginning. In some of these places, the industry's practices are environmental racism, where people of color are targeted for and experience the highest rates of environmental degradation. For example, Kern County, California, where just over half of the population is Latinx, has the highest oil and gas well count of any county in the United States—77,497 active wells (Kelso 2015)—and plans to build 43,000 more (Herr 2021). In California, school districts with greater Latinx and nonwhite student enrollment are more likely to contain more oil and gas drilling and well stimulation than predominantly white districts (Ferrar 2014). California students attending school within one mile of oil and gas wells are 79.6 percent nonwhite (Ferrar 2014).

However, fracking also occurs in predominantly white communities (e.g., areas of North Dakota and Pennsylvania). Places of extraction even creep near the upper class, as when a fracking water tower was to be constructed near the home of Rex Tillerson, Exxon's former CEO. Tillerson joined a lawsuit in 2013 to prevent it (Gilbert 2014). The geographical expansion of the experience of fossil fuel extraction multiplies globally through the climate-changing effects of burning these fuels. Even people without extraction occurring near them experience the negative effects of extraction in the form of a changing climate.

The geographical expansiveness of fossil fuel extraction is an important factor in community resistance to extractivist development and domination, what Willow (2019) calls "extrACTIVISM" and Klein (2014) calls "blockadia." Blockadia refers to the ever more interconnected resistance to environmental and social injustices that is facilitated by the geographic scope of socially and environmentally damaging fossil fuel infrastructures, new technologies of communication, and increasing recognition, on the part of social movements, of their shared targets and goals (Klein 2014). The fossil fuel industry's expansiveness gives protesters many targets. Protesters connect their resistance to sites of extraction and refining, climate change, pipelines, natural gas wells, and fossil fuel transportation by rail, truck, and ship. These elements can be seen as part of a vast fossil fuel industry web, connecting affected communities. The reality of the breadth of fossil fuel-related injustice and protest moves opponents toward a justice perspective, from not in my back yard (NIMBY) to never in or under anyone's backyard (NIABY or NIUABY).[10] This development creates novel conditions for resistance to extreme energy extraction.[11]

Resistance to extreme energy not only has a global character, but also many local sites. As with the substances responsible for environmental injustice (e.g., polychlorinated biphenyls [PCBs] and lead), the substances in question, hydrocarbons, are abundant in society and everyday life. Their extraction with today's intensive techniques worsens conditions in already sacrificed zones—areas of the world whose sacrifice, in terms of environmental and social destruction, enables the lifestyle of others (Bell 2014; Fox 1999; Scott 2010)—and threatens spaces previously little altered by humans, such as the Athabasca River basin in Alberta, Canada. For all of these reasons, extreme energy extraction holds potential for uniting diverse constituencies in resistance. It presents an opportunity for studying how people work across spaces, identities, and issues to defend their communities in the face of climate crisis.

The Climate Justice Movement

As Escobar (1992), Kelley (2002), and Pellow (2014) have all emphasized, the task of social movements is to "construct collective imaginaries capable of orienting social and political action" and "alternative visions of democracy, economy, and society" (Escobar 1992: 41, 22). My interviewees are all doing this, to different extents and with different visions. The work of all interviewees, and the organizations they are a part of, has positive implications for the climate and for social justice. Therefore, I see them as part of the climate justice movement.

The climate justice movement, a broad coalition of organizations and individuals, emerged in the 2000s as a fracturing of the climate movement—civil society organizations that had spent much of their energy lobbying in the context of the yearly United Nations climate change negotiations. In this period, the climate movement had suffered major defeats when the Copenhagen climate negotiations and US cap-and-trade legislation both failed in 2009. In Copenhagen, instead of a global climate treaty, activists walked away with a two-page "accord." The United Nations' lack of progress in addressing climate change led members of civil society to begin to change how they framed their movement from a climate movement to a climate justice movement (della Porta and Parks 2014).

In contrast to "climate," "climate justice," as a frame—a way that social movements describe the problems they face, the solutions they propose, and why people should get involved (Goffman 1974; Snow et al. 1986)—builds bridges between people and movements primarily focused on the environment and those primarily focused on social issues like social justice and anti-war (della Porta and Parks 2014). It is a way of understanding climate change as a crisis of societal relationships with nature, while comprehending that the solution is comprehensive system change (Bedall and Görg 2014).

The climate crisis will not be resolved solely with a focus on technical goals, like numbers, and reforming institutions. Technical and reform-oriented goals are core elements of strategies put forth by mainstream elements of the climate movement. For the climate justice movement, addressing climate crisis requires radical social change, defined as "a deep transformation of a society (or other entity such as a community, region, or the whole world) in the direction of greater economic [as well as racial, gender, and sexual] equality and political participation, accomplished by the actions of a strong and diverse popular movement" (J. Foran 2016). The heart of the movement, therefore, is in the more radical branches of the climate movement which engage in direct action, are explicitly anti-capitalist, and focus on solidarity.

While the climate justice movement first began mobilizing around United Nations negotiations, after Copenhagen it increasingly set its sights on local mobilization. One of its most common tactics is to organize days of action where communities around the world take part and share photos of their actions with each other. One organization that often publicizes these days and posts the photos that result is 350.org. The movement, however, has also been mobilizing more centralized large-scale actions such as the four-hundred-thousand-person 2014 People's Climate March in New York City. These actions prioritize movement building. "To Change Everything, We Need Everyone," as the slogan for the 2014 People's Climate March says, communicates the movement's desire to broaden its base through inclusivity. In 2016, there were major direct actions to disrupt fossil fuel infrastructure in twelve countries on six continents as part of an action called "Break Free from Fossil Fuels." In April 2017, the People's Climate Mobilization in Washington, D.C., with two hundred thousand participants, was even clearer in its commitment to intersectional thinking and action on climate justice. Its website tied climate change to jobs, justice, resistance, and creating a different future. The page read:

We Resist.
We Build.
We Rise.
March for jobs, justice and the climate. (People's Climate Movement
2017)

The website billed the event as "a powerful mobilization to unite all of our movements" (People's Climate Movement 2017).

Unlike the United Nations (UN) negotiations, these local actions are in line with the value the movement places on democracy and horizontal organizing. As Müller and Walk (2014) point out, the UN negotiations are inherently undemocratic, allowing a small number of negotiators from around the world to make decisions (or fail to make decisions) about climate change. The civil society organizations that have a place (though increasingly restricted) at the UN negotiating table are made up of professionalized and resource-rich people who are not representative of Earth citizens. The spaces, policies, and organizing at the negotiations exclude Indigenous and marginalized communities (Grosse and Mark 2020). At the international and EU level, the NGOs with influence are those that see eye to eye with governments (Müeller and Walk 2014:39)—there is no room for contestation, an essential element of democracy. There is also little room for understanding many aspects of climate policy because of the excessively technical language, which even experts find "barely comprehensible" (Müeller and Walk 2014:39; see also Grosse and Mark 2020). These features of the UN system have contributed to the movement's refocus on local campaigns.

The climate justice movement's turn to the local, where women are overrepresented in grassroots organizing, means that more women are shaping the movement. Women's participation, by embedding feminine ethos and practices in relationships with other activists and groups, is contributing to the movement's focus on movement building and intersectional justice—justice that addresses the root causes of many forms of oppression.

The local campaigns that activists are mobilizing prioritize issues as wide ranging as climate justice, pipelines, fossil fuel divestment, oil trains, anti-extraction, and renewable energy. Motivations are also varied. For example, the fight against the Keystone XL pipeline had spokespeople and participants including the Cowboy Indian Alliance, a coalition of ranchers, farmers, and tribes along the pipeline route working to defend land, water, and property; climate justice activists; and faith communities. As one reporter wrote: "It's one of the distinctive features of the anti-pipeline movement that no two activists are fighting for quite the same thing" (Brown 2016), except to stop the pipeline.

Across Turtle Island (North America), Native-led fossil fuel resistance efforts have galvanized impressive coalitions in recent years, with tribes and Indigenous organizations such as Indigenous Environmental Network, Affiliated Tribes of Northwest Indians, Honor the Earth, and Idle No More playing key roles. Through their centering of the land and water, relationships with relatives, and the well-being of future generations, Native Nations' assertions of their sovereignty and treaty rights have transformed the skills and outlooks of water protectors of diverse identities. Indigenous peoples are on the leading edge of protecting the earth, its human and more-than-human communities, and offering paths forward for sustainable relationships, drawing on Indigenous knowledge. This is evident in the successful resistance of First Nations in Canada, Native Nations in the United States, and non-Native groups against tar sands pipelines (see LaDuke 2016 on many of these struggles), including the Keystone XL pipeline (see Black et al. 2014a), Northern Gateway Pipeline (see Bowles and Veltmeyer 2014), and Energy East Pipeline. It is clear in the Standing Rock Sioux Tribe's ongoing efforts to protect water from the fracked oil Dakota Access Pipeline (see Estes 2019; Estes and Dhillon 2019; Gilio-Whitaker 2019); Wet'suwet'en assertions of rights against the Coastal GasLink fracked natural gas pipeline; and the Anishinaabe mobilization to stop Line 3 (see LaDuke 2020).[12] These struggles build on a long history of Native and non-Native relationships, alliances, and coalition building around land rights, environmental, and social justice issues (see Davis 2010; Grossman 2017).

While these movements are first and foremost about sovereignty, the rights of Native Nations to self-govern and control their land, and treaty rights that are guaranteed by Article VI of the US Constitution, they also advance the goals of environmental and climate justice. These movements demand that people affected by environmental hazards be involved in decision making, that dangerous infrastructure not be allowed to contaminate the land, cultural sites, and water, and that oil be kept in the ground.

In light of the climate justice movement's increasing visibility and power—the 2014 People's Climate March was the largest climate march in history and Standing Rock was the largest gathering of Native nations in hundreds of years—and the increasing severity of climate crisis, there is much work to do to improve the efficacy of the climate justice movement and the scholarship that illuminates its work. Most existing academic research analyzes the climate justice movement at the annual UN negotiations (e.g., Bedall and Görg 2014; De Lucia 2014; della Porta and Parks 2014; Foran and Widick 2013; Müller and Walk 2014). Attention to the local could therefore broaden understanding of the movement's manifestations in different contexts and deepen understanding of non-UN targets, tactics, and strategies.

Related to this, research in local contexts can clarify how the movement works within and across local and global scales, organizations, and issues—how its members build coalitions. Bond (2014) and Harlan et al. (2015) both underline this as an important area for research, calling on social scientists to broaden understanding of the barriers the climate justice movement faces, coalition partners that will be key for its success, and opportunities and challenges for transnational organizing against fossil fuel corporations. To this end, I examine how people and groups within the climate justice movement work together to resist local extreme energy extraction and to imagine and create a just and sustainable world.

CONCLUSION

The contradiction of climate crisis and continued extreme energy extraction is a symptom of the dissolution of just, equitable, and sustainable relationships among people and between people and the more-than-human world. In the chapters that follow, I trace how activists are focusing on relationships to address this crisis.

This chapter sets the stage for subsequent chapters by grounding the reader in broader contexts of climate crisis, extreme energy extraction, and the climate justice movement—conditions that shape the lives of activists I interviewed. The next chapter explores the place-based context of my research.

CHAPTER 2

The Organizing Landscape

Research Context

Place informs how people experience and learn about the world (Feld and Basso 1996). It also informs the siting of energy extraction, transportation, processing, and consumption, as well as climate change impacts for humans, other living beings, and the physical environment. As Escobar writes, "place, body, and environment integrate with each other" (2001:143).

My research is grounded in communities in two western American states: Idaho and California.[1] They are places where I am personally and politically embedded in long-term relationships with people and environments. They contrast in dominant political ideologies, strength of environmental movements, climate change policy, and histories of energy extraction. California has been a leading oil producer for over a century, whereas Idaho is a fossil fuel frontier—a place that has never, until 2009, had any commercially viable fossil fuel extraction. As I explain in chapter 8, this contrast, combined with other features of these places, affects how activists organize against extreme energy extraction.

The national context also affects the movements and communities I study, for they exist within the social, cultural, political, and economic milieus fueling the climate crisis. The United States is both the global leader of hydrocarbon production (since 2013) and also the highest historical contributor to carbon dioxide emissions (US Energy Information Administration 2016a; World Resources Institute 2014). In interviewee Cass Davis's words, "Being a poor peasant, there's no place I can cause indigestion better than in the belly of the beast." I anticipate activists' insights in these locations within the United States, and I intend my analysis to be useful for individuals and communities working within and between similarly situated places.

OIL AND CLIMATE ACTION? CALIFORNIA

Despite its progressive climate policies and green image (see Megerian 2015), California has a long history of oil production. It was the third-largest US producer of crude oil historically until 2017 (U.S. Energy Information Administration 2021b).[2] People (mostly men) have been drilling for oil and gas in California since the mid-1800s.[3] The first boom occurred in the mid-1860s when sixty-five oil companies drilled from Humboldt Bay to Ventura (Division of Oil, Gas & Geothermal Resources 2013b). In response, the Department of Petroleum and Gas was established in 1915 to enforce industry-related regulations legislated since 1903. It was later renamed the Division of Oil, Gas & Geothermal Resources (DOGGR) and, as of 2020, the California Geologic Energy Management Division (CalGEM). Summerland, in Santa Barbara County, was the site of the world's first offshore drilling in 1896. In 1969, the first major offshore oil spill in the world occurred in the Santa Barbara Channel; it took the well operator ten days to stop the flow of oil and gas (Molotch 1970). The event helped prompt the first Earth Day in 1970, the creation of many environmental organizations and the University of California Santa Barbara's environmental studies program, and the adoption of environmental and extraction regulations. The first comprehensive report on the state's offshore oil and gas seeps was published two years later, in 1971. Since then, DOGGR has been working to clean up improperly abandoned oil infrastructure throughout the state, and now plays a role in the regulation of unconventional techniques such as fracking.

In 2013, California governor Jerry Brown signed Senate Bill 4 (SB4) that mandates regulations, which went into effect in July 2015, for three kinds of well stimulation: fracking, acid fracturing, and acid matrix stimulation. Acid fracturing uses acid to fracture rock. Acid matrix stimulation applies acid to the well or geologic formation at lower pressures than is necessary for fracturing. SB4 requires oil and gas operators to notify DOGGR before using any of the three well-stimulation activities, expand monitoring and reporting of water use and quality, conduct broad analysis of potential engineering and seismic impacts of extraction, and disclose chemicals used. The *Los Angeles Times* called the

FIGURE 1. Protest sign featuring former California governor Jerry Brown. March for Real Climate Leadership, Oakland, California, February 7, 2015. Photo by the author.

regulations the "toughest-in-the-nation fracking rules" (Cart 2015b). The state's ambitions to limit dirty energy production, however, stop at regulations. After SB4 was first approved, Governor Brown consistently ignored calls by environmental organizations, grassroots groups, and concerned residents to ban fracking.

On March 15, 2014, four thousand marched on the Capitol in Sacramento at the Don't Frack California Rally. On February 7, 2015, eight thousand participated in the March for Real Climate Leadership in Oakland, one of the largest anti-fracking demonstrations in US history. In May 2016, in the aftermath of the largest methane gas leak in US history in the Los Angeles neighborhood of Porter Ranch, two thousand people marched in Los Angeles to "Break Free from Fossil Fuels." The satirical protest sign of former governor Jerry Brown (figure 1), present at many of these protests, sums up activists' frustration. Many do not think the toughest-in-the-nation fracking rules are very good. Hollin Kretzmann of the Center for Biological Diversity said, "The state's weak fracking rules focus on notification and do almost nothing to safeguard California's air, water or public health" (Cart 2015b). Other concerns include regulators' capacity to enforce the rules and the quality of DOGGR's environmental impact report on fracking (Cart

2015b). Concerns about enforcement capacity stem from an internal audit by DOGGR finding that, among other violations of policy, since 2007, most oil projects in the Los Angeles area have not been subject to a required annual review (Cart 2015a). In 2021, Governor Newsom announced a ban on new California fracking permits starting in 2024. He also asked regulators to consider phasing out all oil production in the state by 2045.

In addition to state-level campaigns, individual counties and cities throughout California have marshaled opposition to fracking, acidization, and another intensive technique, cyclic steam injection. Cyclic steam injection uses large quantities of water to create steam that is injected at high pressure to melt tar-like oil. San Benito, Mendocino, and Santa Barbara Counties all gathered enough signatures for ballot measures to ban fracking in the 2014 mid-term elections. San Benito gathered forty-one hundred signatures, Mendocino gathered six thousand, and Santa Barbara gathered twenty thousand. Mendocino and San Benito voters passed the ballot measures.[4] Both counties had little or no oil and gas production. In Santa Barbara, the sixth-largest oil producing county in California in 2013 (Division of Oil, Gas & Geothermal Resources 2013a), with a century-old oil presence, and where over six million oil corporation dollars funded an opposition campaign, the ballot measure did not pass. In 2016 ballot measures, Butte County and Monterey County—California's fourth-largest oil producing county at the time (Division of Oil, Gas & Geothermal Resources 2013a)—successfully passed bans. In 2014 and 2016 respectively, Santa Cruz and Alameda counties banned fracking through approval by county leaders. Of all of these efforts (see map 1), the quantity of signatures and the campaign spending and organizing on both sides make Santa Barbara the largest county-level electoral mobilization to ban fracking in California state history.

A grassroots chapter of the international group 350.org, 350 Santa Barbara (350SB), spearheaded the Santa Barbara mobilization.[5] Though Santa Barbara has a number of environmental groups that have been doing good work for decades (the Community Environmental Council (CEC) was formed in 1970 and the Environmental Defense Center (EDC) was formed in 1977), at the time of 350SB's founding in January 2013, few of these groups were dedicated entirely to climate change or engaged in protest. These groups also have long-term relationships with donors in the Santa Barbara community and local government officials, something that 350SB members saw as making them inclined to work within more conservative venues for creating environmental

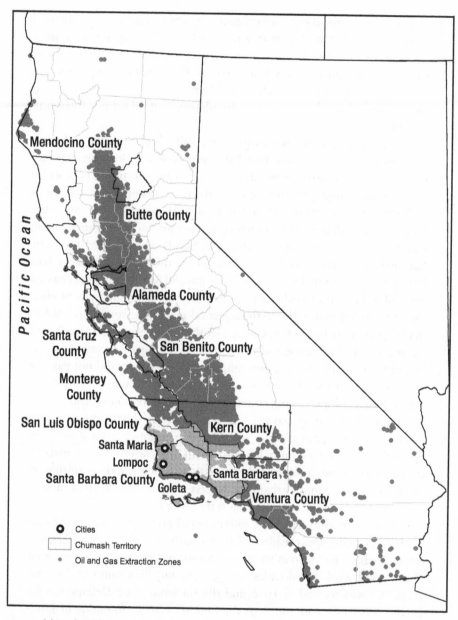

MAP 1. Map of California research sites, counties surrounding Santa Barbara County, counties with fracking bans, and oil and gas extraction zones. Map by Will Matuska.

FIGURE 2. Protest against the Keystone XL Pipeline, Santa Barbara, California, September 2013. Photo courtesy of Hunter Grosse.

change (legal, consumer choice, public hearings), rather than engaging in grassroots community organizing, protests, and ballot measures. In fact, leaders of these and other longtime environmental groups in Santa Barbara explicitly advised against 350SB organizing the 2014 ballot measure. When the measure failed, these groups communicated an "I told you so" sentiment to 350SB and expressed fear that the failure would damage the environmental movement in Santa Barbara and elsewhere.[6]

In line with 350.org's message to bring carbon dioxide concentrations in the earth's atmosphere down from their current level of over 400 parts per million (ppm) to 350 ppm, a safe level for continued life on earth, 350SB is focused on climate change. The group has hosted rallies for renewable energy, educated the community about climate change, and urged city and county institutions to divest from fossil fuels. The group's ranks swelled when, in September 2013, members constructed a ninety-foot-long inflatable replica of the Keystone XL Pipeline, and with about one hundred people in attendance, marched it through downtown Santa Barbara to the beach (see figure 2).

The next campaign to spark the interest of 350SB was fracking. In March 2014, faced with oil industry plans for seven thousand seven hundred new unconventional oil wells in northern Santa Barbara County, two key women organizers, Becca Claassen and Katie Davis, drafted an anti-fracking ballot measure, modeled on that of San Benito. Activists had to collect 13,200 signatures to put the measure on the November 2014 ballot. A coalition organization that 350SB put together called the

Santa Barbara County Water Guardians gathered 20,000 signatures in three weeks in April.

In the following months, one thousand volunteers worked within the context of the "Yes on Measure P" campaign to educate voters (see chapter 7 for a detailed account of Measure P). Californians for Energy Independence, with majority funding by Chevron and Aera Energy, spent over six million dollars in the "No on Measure P" opposition campaign. "Yes on Measure P" raised over four hundred thousand dollars. In the end, 61 percent of voters voted no and 39 percent voted yes in a midterm election with especially low voter turnout. Nationwide, it was the lowest voter turnout in seventy-two years (Alter 2014) and in Santa Barbara, only 56 percent of registered voters cast votes on Measure P.[7]

Since 2014, people who worked on Measure P have protested other extreme energy issues. In 2015 and 2016, they sent public comments, attended community meetings, and protested at hearings to stop Phillips 66's expansion of its rail spur in Nipomo, just north of Santa Barbara County in San Luis Obispo County, with six hundred people protesting at a hearing in February 2016. This expansion would have increased oil transportation on railways and through communities along the California coast, part of a larger trend of oil by rail sweeping the country. In March 2017, the San Luis Obispo County Board of Supervisors rejected the project.

In 2016, the other most notable campaign that engaged Santa Barbara climate justice activists was Standing Rock. Many 350SB members collaborated with local Indigenous Chumash leaders to organize the Santa Barbara Standing Rock Coalition, which hosted marches and divestment protests, where attendees publicly divested from banks funding the Dakota Access Pipeline. This campaign quickly combined with rallies and marches in protest of the November 2016 election of Donald Trump. Trump held shares in Energy Transfer Partners, the Dakota Access Pipeline owner, until summer 2016 and had made statements in support of Dakota Access Pipeline during his campaign, ultimately authorizing its construction by executive order just after taking office.

CLIMATE DENIERS IN A FOSSIL FUEL FRONTIER: IDAHO

While fossil fuel production has long been a part of Santa Barbara County and California, it has only recently begun in Idaho. In 2016, the hottest year ever recorded at that point, three of Idaho's four members of the US Congress were climate deniers and state lawmakers dropped climate change from K–12 education standards (Ellingboe and Koronowski 2016; Kruesi 2017).[8] Climate denial characterizes Idaho's leaders, the people who are making decisions about fossil fuel development, but not the public; 75 percent believe in climate change, compared to 82 percent in California (Krosnick 2013).

Idaho is a rural state with approximately 1.78 million residents in 2019.[9] While industry has been searching for oil and gas since 1903, and drilled about 145 wells through 1990, until recently "the story of oil and gas exploration in Idaho [was] an ongoing saga of near successes and shattered expectations" (McLeod 1993). No commercial discoveries were made. Since 2009, however, fifteen natural gas wells have been drilled in Payette County, in southwest Idaho, and three permitted in Bonneville County, in southeastern Idaho (Idaho Department of Lands 2017) (see map 2). In 2017, nine were producing, up from one in 2015, and six awaited pipelines. From the beginning, residents have feared that production will use fracking.

Resident suspicions are based on anecdotes and a distrust of government and industry. One community member, in a tour of the local power plant, had the CEO of Snake River Oil and Gas, Richard Brown, tell him that gas companies planned to frack, a statement overheard by other community members. Bridge Energy, a company that has since gone bankrupt, is on the record using the term *mini-fracking* to describe well-stimulation operations in Idaho (see Prentice 2011a). Residents fear fracking despite government, industry, and environmental lobby groups telling them that no fracking occurs in Idaho, and that they have a conventional gas play, meaning gas easily flows into the well and the surrounding rocks therefore do not need to be fractured. Activists, however, use the term *fracking* to describe what they are fighting against and believe that it will occur. As is the case in Santa Barbara, where activists would like to have cyclic steam injection included, and therefore prohibited, within the concept of fracking, perhaps as "steam fracking," how people understand and employ the concept of fracking may not match what is technically considered fracking by the oil and gas industry.

Oil and gas development in Idaho is under control of the Oil and Gas Conservation Commission, which, until 2013, was the five members of the Idaho Land Board. That year, Governor C. L. "Butch" Otter and the Idaho Legislature changed the composition of the Commission to five individuals appointed by the governor and approved by the Senate. Many of the regulations for current oil and gas development in Idaho were developed in 2014 and approved by the Idaho State Legislature in 2015.

Though the regulations improved existing environmental protections, critics such as Alma Hasse and Tina Fisher, cofounders of Idaho Residents Against Gas Extraction (IRAGE), say they are insufficient (Smith 2014a). Alma, Tina, and a small group of concerned citizens in Payette County and Gem County (where seismic testing for drilling occurred before 2016) were the first force of opposition to natural gas expansion, arguing for enhanced environmental testing, increased distance between drilling and homes, better preparation of public safety providers, and greater industry transparency. Having gained little traction in local communities while working as IRAGE and spreading messages focused on the environment, in summer 2015, Alma and other residents formed a new nonprofit organization, Citizens Allied for Integrity and Accountability (CAIA). Along with communicating the risks of fracking, CAIA has been sounding the alarm about the possibility that landowners can be in default of their mortgages if they sign an oil and gas lease.[10] CAIA has experienced much greater rates of growth than IRAGE. In chapter 4, I argue that this is a result of their messaging and appeals to common values.

Idaho oil and gas policy is comprised of bills that serve the interests of the gas companies. Idaho House Bill 50 states that Idaho residents can be force-integrated into gas extraction projects by the state if owners of 55 percent of the mineral rights in the surrounding section of land agree to lease. That means 45 percent of mineral rights owners have no say in whether extraction occurs. If the state issues an Integration Order, prompted by an oil and gas company, mineral rights owners in the section have three options.[11] These are outlined by the Idaho Department of Lands (2015): (1) "Participate as a Working Interest Owner," in which mineral owners pay for the extraction infrastructure (hundreds of thousands or millions of dollars split between mineral owners) and receive full royalties, (2) "Elect Non-Consenting Working Interest Owner" in which mineral owners who do not have the money to pay for the infrastructure pay a 300 percent penalty on the cost of the infrastructure and receive a reduced 1/8th royalty, and (3) "Deemed Leased" in which mineral owners refuse to participate, but are forced to lease by the Integration Order. In this case, mineral rights owners receive 1/8th royalty. Two other particularly egregious policies, in CAIA's eyes, are House Bill 464 and Senate Bill 1339. Passed in 2012, House Bill 464 stripped all local control over the siting of wells and removed local capacity to prohibit extraction (essentially a ban on bans that prohibit fracking). Senate Bill

FIGURE 3. Megaload. Photo courtesy of Dave King.

1339, passed in 2016, removed CAIA's ability to intervene in court on behalf of its members (this had been the group's primary legal strategy); drastically shortened the approval process for drilling applications; and gave all power of approval for oil and gas infrastructure to one person, the director of the Idaho Department of Lands. Like many of the other oil and gas bills, Senate Bill 1339 was declared an "emergency" so it could go into effect immediately.[12] Implications of these regulations, and especially House Bill 50's illogical penalties, are central concerns of CAIA. See chapter 4 for a more detailed description of CAIA's struggle with oil and gas legislation.

Natural gas extraction is just one of an increasing number of fossil fuel–related activities that link Idaho—where 12.6 percent of the state is wilderness (the highest proportion in the United States) and there is about one and half times more cattle than people—to a global extractivist economy (see Hormel 2016). From 2011 to 2014, Idaho highways were sites of transportation of tar sands extraction infrastructure that, together with their trailers, were known as megaloads (see figure 3).[13]

Along with a few other ports, the equipment was shipped into the Port of Lewiston, Idaho, the farthest inland port from the Pacific, and then transported via megaload to the Kearl oil sands in Alberta. These two-hundred-foot-long, two-story-building-tall, and two-traffic-lanes-wide trucks attempted to expedite their trip to Alberta by traveling along Highway 12, a federally designated scenic byway, Highway 95, and a handful of other routes (see map 2).[14] On each, protests by residents and tribes slowed the megaloads' progress. Grassroots groups and nonprofits such as Wild Idaho Rising Tide (WIRT), Fighting Goliath, Friends of the Clearwater, and Idaho Rivers United were key resistance organizers. In a powerful action in August 2013, eight members of the Nez Perce Tribal Executive Committee, including chairman Silas Whitman, asserted their sovereignty and were arrested while blockading a megaload on their reservation. In January 2017, a long legal battle concluded with a prohibition of megaloads traveling along Highway 12.[15]

As with Measure P, after 2014, activists who had worked on the megaload campaign began working on other extraction issues. Idaho railways, running along rivers and across lakes, carry oil, tar sands, and coal from northern sites of extraction to the west coast. Thus, oil trains, as in Santa Barbara, became a topic of public concern (Associated Press 2014; Wild Idaho Rising Tide 2015). In Sandpoint, Idaho (see map 2), trains carrying coal from Wyoming, oil from the Bakken Shale in North Dakota, and tar sands from Alberta, Canada, come together and cross Lake Pend Oreille, the largest lake in the Idaho Panhandle and fifth deepest lake in the United States. Helen Yost, the central organizer of WIRT, as well as staff in the Sandpoint offices of the Idaho Conservation League and Lake Pend Oreille Waterkeeper are currently working with Sandpoint residents to raise awareness of, and in WIRT's case, engage in direct action against the trains.

In 2016, again in parallel to Santa Barbara, Idaho activists engaged with Standing Rock. The Nez Perce Tribe, who fought the megaloads, issued a resolution in solidarity with the Standing Rock Sioux Tribe against the Dakota Access Pipeline (Nez Perce Tribal Executive Committee 2016). Alma Hasse, a veteran and board member of CAIA, joined over four thousand other veterans who traveled to Standing Rock in December 2016 to support the water protectors. On December 5, they held a ceremony in which they asked for forgiveness from Native Americans for the crimes of the US military (Taliman 2016).

Idahoans are resisting each of these elements of the global fossil fuel infrastructure—natural gas extraction, tar sands infrastructure, fossil fuel trains and pipelines—with little support from nongovernmental organizations or state and local governments. They work in a political context that is hostile to environmentalism and in a social context where most people are conservative. Interviewee and resident of Kooskia,

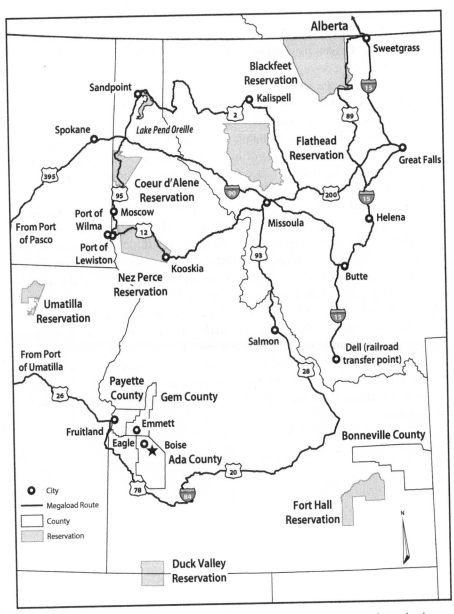

MAP 2. Map of Idaho research sites, counties with natural gas development, and megaload routes. Map by Will Matuska.

Idaho, Lin Laughy explained that for liberals "it's tough sledding around here." In contrast, state level and national groups like Food and Water Watch, the Center for Biological Diversity, and Stand.earth (formerly ForestEthics) have supported Santa Barbara activists. In addition, California's politics are known for being environmentally progressive. Both California Legislators from Santa Barbara's district during 2014, Senator Hannah-Beth Jackson and Assemblyman Das Williams, endorsed Measure P, and in the case of Williams, made the Santa Barbara fracking ban ballot measure a central part of his own campaign. As of March 2015, the California legislature was 64 percent Democratic while the Idaho legislature was 19 percent Democratic (Ballotpedia 2014).

Idaho and California, therefore, are diverse cases for the study of resistance to extreme energy extraction. In the chapters that follow, I show how these settings shape the nature of resistance. While activists' methods for working across lines differ depending on context, activists across my research sites prioritize coalition building as one of the most important parts of their work.

In the next chapter, I introduce the reader to some of the key activists leading resistance to natural gas in Idaho. Their journeys into activism and how they understand what the word "activist" means illustrate the importance of relationships for trust within communities and for talking across lines, the primary tactic of these activists. On the flip side, lack of relationships can fuel mistrust and injustice and inhibit coalition building.

CHAPTER 3

Idaho Part 1

*Talking across Political Lines
by Building Relationships*

In this chapter, I delve into what it means to talk across political lines, typically those that demarcate conservative and liberal ideologies. I draw on what I learned from folks resisting fracking in western Idaho, from how they talked with their fellow group members, and, as I observed in my time living with them, how they spoke to people like the dishwasher repairman, or the postal worker they encountered in their daily lives.

An important condition for talking across lines is relationships and trust. When these exist, there is potential for people to hear each other and to work together. When these do not exist, building coalitions is challenging. Understanding how trust is built or inhibited depends on understanding activists' lived experiences, values, and the social setting in their community. Therefore, I begin my account of activism in southwest Idaho with stories of interviewees' journeys into activism. These experiences provide important grounding for why activists have needed to develop the tactic of talking across lines and illustrate the centrality of relationships to building resistance to extreme energy extraction.

I open with a story related to the 2016 US presidential election, a series of events characterizing the context in which interviewees expressed and developed their views on politics and organizing. This account illustrates two points. First, it displays the polarizing nature of labels in political rhetoric and, consequently, the need to talk across lines by drawing on common values. Secondly, it evidences the promise of one of this book's core themes: relational organizing.

LUNCH CONVERSATIONS ON SOCIALISM

On February 18, 2016, Bernie Sanders, self-proclaimed socialist competitor for the Democratic candidate for the 2016 US presidential election sent me a campaign email. In it, he highlighted how his campaign was succeeding because of support from working people, not elites. The email explained that elites, unable to drive Sanders's campaign, had lashed out in the lead-up to the February 20 caucus in Nevada, "not just at Bernie, but at you." A bullet point in the email read, "Bill Clinton compared supporters powering our political revolution—people like you—to the Tea Party. The Tea Party!"

Just two months prior, in December 2015, I had been spending a lot of time with folks sympathetic to Tea Party ideas. With one, seventy-six-year-old Dottie Hawthorne, I had even had a productive conversation about Bernie Sanders. In our interview over tuna sandwiches in Dottie's living room, Dottie had described socialism and communism as close to one another and "frightening, so frightening." "You cannot take from the rich and give to the poor. [. . .] You need to give the poor incentives to grow and to earn and to be proud of themselves. [. . .] It's against my way of thinking common sense." Later, disgusted with both "sides of the aisle" of US politics, Dottie asked: As far as our people who are going to run for president, my goodness, which one would you choose, if you, no political affiliation, just which of the candidates would you choose right now?

Corrie: Bernie.

Dottie: And why?

Corrie: Because he's a socialist.

Dottie: And you think socialism is good?

Corrie: Yeah.

Dottie: I guess I need to know why you think that.

We proceeded to discuss socialism, welfare, and conservatism. Though Dottie was not convinced at the end our conversation that Bernie was "who we can count on to see that [taxation] is done right," she did say, "Eh, food for thought, huh, very interesting, I have not read very much about Bernie Sanders."

Here, in this brief statement, I see a kernel of hope. Dottie, a seventy-six-year-old conservative rancher, and me, a twenty-five-year-old woman who had spent her last four years under the mentorship of radical Marxist sociologists in California—the bastion of liberalism in the United States—were able to have a calm, respectful conversation about socialism. We did not shoot each other down, but listened, and, based on listening, offered examples and perspectives gleaned from our lived experiences. We shared our views with each other. We talked across lines. Part of talking across our lines of difference was agreeing to disagree, being able to sit with that disagreement while enjoying each other's company, and appreciating each other's contribution to our common cause—the fight against fracking.

This is the kind of conversation I see becoming more possible because of the organizing practices of Citizens Allied for Integrity and Accountability (CAIA), a grassroots nonprofit organization dedicated to government and corporate integrity and accountability and, therefore, the fight against fracking. Dottie's and my conversation was one that necessitated a relationship of trust, which usually, and especially when two people hold opposing views, requires time.

Bernie's email to me, had Dottie read it, would have undone any progress she and I made in communicating to each other. It was an email to bind *certain* people together *against* others. In this case, others were Tea Partiers and the Democratic establishment represented by Bill and Hillary Clinton (Bernie's competitor). This type of writing denies the possibility of conservative people working for progressive change, people like interviewees Jan and Wayne.

Jan and Wayne were former leaders of a local Tea Party group and core members of CAIA. They stressed the local character of their Tea Party group, and the diversity of Tea Parties more generally, identifying limited government control and spending as the one unifying theme. They argued that their political affiliation had nothing to do with their stance on oil and gas, a statement that could come as a surprise to someone who supposes that Tea Party equals conservative equals pro fossil fuels. As Jan explained, "People are usually surprised because we feel the way we feel about oil and gas [and] are Republicans or conservative. It's like you can't be that way. And it's like, 'What do you mean?' If you are thinking, really thinking about the issue and you don't like it, it has nothing to do with what your beliefs are politically." Wayne's language was more colorful: "Politics does not play a role in this big picture [of oil and gas]. I don't give a rat's behind about your political affiliation; I want to stop these assholes from what they are doing." Jan and Wayne were able to organize with self-identified Democratic anti-fracking activists because the group expressly rejected stereotypes embedded within political labels. In fact, the group worked to avoid use of labels entirely, in their messaging and internal discussion, something I delve into in the next chapter. What is the effect, then, of pitting Bernie supporters against Tea Partiers? Might it be more productive for building progressive change, for building climate justice, to think and communicate in unifying, rather than dividing terms?

Communicating in unifying terms was exactly what CAIA had developed as its primary organizing strategy. CAIA members did not come up with this strategy out of the blue, however. Talking across lines was a hard-won strategy developed to meet the needs of activists organizing amid small, politically red farming towns (approximately 75 percent of voters in this area voted for Donald Trump in the 2016 election), activists who had started out as people who did not think of themselves as activists.

EXTREMISTS, HIPPIES, AND BIRKENSTOCKS: AMBIVALENCE ABOUT "ACTIVIST"

Like fracking opponents in other parts of the country (see Willow 2019), most CAIA members did not think of themselves as environmentalists or activists. However, in the course of our interviews, many explored the activist label as they addressed my question, *How do you describe yourself?* which I typically followed with examples, *activist? organizer? concerned citizen?* A few initially hesitated about using "activist," but then later came to the conclusion that actually, they were activists, because activists are not limited to the stereotypical image of an extreme, protesting, Birkenstock-wearing hippie, to borrow imagery from multiple interviewees. Alma Hasse, who had gone to jail for her anti-fracking efforts, had dramatically changed her views about activists and environmentalists. As she explained, "I used to—I am ashamed to say this, Corrie, don't hold it against me—I used to say environmentalists were Birkenstock-wearing hippies with too much time on their hands. Now I know they don't wear Birkenstocks, don't have enough time, and think, 'where would we be without them?' We are lucky to have them." Alma's admission is emblematic of how interviewees' conceptions of activists transformed over time. Most thought of themselves as regular people until an event jeopardized their sense of safety. Still thinking themselves regular people, they worked to protect themselves and quickly found that they were no longer viewed as regular by their neighbors. All of a sudden, they were perceived as in league with Birkenstock-wearing hippies, even though none of them wore Birkenstocks or matched any stereotypes of hippies. This led interviewees to reassess what they thought of activists and to understand their own activism against natural gas as important, no matter what their neighbors thought.

Luke and Brynna Smith, a couple in their early thirties who were part of a legal effort to stop natural gas next to their property, described themselves as concerned homeowners and parents. They told me that activists were viewed with suspicion in their community, as "squeaky wheels," Brynna said. Luke explained, "If you've been labeled an extremist or an activist you are just [. . .] not wanting to be listened to." According to Luke, viewing activists as extremists was a sharp contrast to how locals viewed the natural gas company representatives. People trusted the company and believed their assurances that everything would be OK. Neighbors cited the nice suits and respectful manner of the landmen, men employed by the natural gas company to knock on doors and ask landowners to sign leases for natural gas, as evidence that they were trustworthy. In this area of Idaho, businessmen were respected, while environmentalists, who were typically women, were not.

Luke and Brynna originally had no opinion on oil and gas. They started researching "both sides" of the issue when a land man brought a gas lease to their house. Their investigation of the issue began online, cross checking everything on the documents that the gas company gave them. A couple of months later, they realized that the proposed well would be constructed right next to their property line and children's swing set. From their research, they concluded that the gas company was not being a responsible steward and that having a well next to their house could have negative health impacts on their four young children. They began writing letters to politicians and initially felt like they were the only ones in the county that were questioning the drilling. They felt isolated. Neighbors told Luke and Brynna that they should be careful speaking out about the proposed well. They explained, "You don't want them to start, you know, totally discounting what you say because you have been pushing the issue," said Brynna. She explained: "So that right there spoke volumes to me about the way people view anyone who really stands up for what they think is right, you know. So what, can I write like three letters to the newspaper or something, what's my limit here before you start thinking I am crazy?" The threat of a well next to their home made the Smiths aware of the importance of speaking out. They began to value the work of activists like Alma and started speaking out themselves.

Discussing the way that gas production processes worked, where "unless you object you are considered [. . .] consenting," Luke said, "You almost have to be an activist just to stop something or let people know, 'Hey, we need to rethink this or look this over before we continue on,' because a simple mistake or simple yes [. . .] can completely wipe this town out." Luke thought it was silly that "either [. . .] your kids better be drinking oil when they eat their dinner or you are a complete activist and you have dreads and everything else, it's one or the other is how it's viewed." Both he and Brynna were interested in ensuring that things in their community were done responsibly and they did not think that was extreme. The categories of pro-oil and extreme activist are too constrained for people like Luke and Brynna and the many other concerned individuals who identify with neither of these categories.

As we continued our conversation, Brynna talked about how most people who think activists are extreme are not willing to "stand up for anything ever" and how she and Luke had been guilty of that. Luke interjected: "We used to call [activists] patriots," the people who "freaked out, [got] people together, [and] did something" to address taxation without representation before the Revolutionary War. Encapsulated in this discussion on what it means to be an activist is Luke and Brynna's larger journey from never having engaged in politics to having their names on the lawsuit against Alta Mesa (the gas developer in Idaho) and considering organizing a petition for enhanced oil and gas safety regulations. Their experience resisting oil and gas had not only changed their view of the industry, but also of the social forces that speak out in communities.

Ambivalence, being "on the fence" about activism, was common. Local media routinely calls Alma an "Anti-Fracking Activist" (e.g., *Boise Weekly, Idaho Statesman, Argus Observer, 580 KIDO*). Yet Alma does not consider herself an activist: "I don't really, and I haven't really ever thought of myself as an activist per se, you know; there is a need and nobody, I didn't see anybody stepping forward to fill it, so here I am [laughs]." So what was the need that spurred Alma to action?

As Alma liked to say, up until she moved to Idaho, her "head had been firmly planted in the sand" and her "rose colored glasses firmly affixed" to her face. When she lived in California, before coming to Idaho, she viewed herself as a conservative, and both her and her husband Jim had all the material things they wanted. Alma had fulfilled the many homemaker duties of a middle-class family concerned with presentation. "I used to dust the top of my fridgerator on a weekly basis," Alma would tell me, as she apologized for what she perceived as a state of disarray in her home, each time I arrived for fieldwork. "Now, I hardly even have time to do the dishes, Jim often steps up to do them." In her sweatpants during a day of computer work, before she quickly changed for a meeting and rushed out the door, Alma would say, "It's a good thing I don't put on makeup anymore." The change from a life of more traditional gender roles and related concerns about presentation of self and home to one of constant resistance to oil and gas—"givin' them hell," as Alma says—paralleled the change in her view of environmentalists from Birkenstock-wearing hippies to invaluable members of civil society. In sum, Alma's priorities had changed significantly since 2006 when she realized she had a CAFO (confined animal feeding operation) as a neighbor. From that point on, activist work filled her time.

Her CAFO neighbor "officially smashed" her rose-colored glasses. Alma asked him to stop leaving piles of cattle feed on the road, fallen from his overloaded trucks. In response, he scooped one up and dumped it on the corner of her property. Alma gathered the sample of cattle feed in a neighbor's borrowed canning jar and had it tested by Analytical Laboratories of Boise. It "had some of the highest levels of E. coli [the tester] had seen." The tester "told me to call the school and have them not pick the kids up there because the kids should have no contact with this stuff," Alma explained. Alma went on to work against unjust CAFO policies at the state level, forming a nonprofit and revealing CAFO non-compliance with state and federal laws (Ogburn 2011). This work parallels CAIA's work against oil and gas. Both fight industry-friendly laws that strip away public process and local control over environmentally risky, and, in climate terms, catastrophic businesses. Both cows and leaking natural gas pipes are tremendous sources of methane emissions, not to mention the carbon dioxide emissions resulting from burning natural gas or transporting cows and beef in our sprawling food system.

Alma's husband, Jim, experienced a similarly intense event that spurred him to become an activist, though he also does not call himself one. On Thursday October 9, 2014, Alma sat with about six other members of the public observing a Payette County Planning and Zoning Commission public hearing on oil and gas. Speaking from the front of the fluorescently lit, unadorned, and nearly empty meeting room, a commissioner accused Alma of presenting false information in her testimony at a previous hearing. The present hearing had entered the deliberation stage, when the public is not allowed to speak. Alma demanded a point of order, asking to know the source that contradicted her statement.

She had previously stated in public comment that Santa Barbara had an ordinance prohibiting the transportation of gas by railway. The commissioner said he called the city's zoning commission and they denied this. In response to Alma's request for information, the commissioners asked her to leave. She would not—they were in a public meeting, she said—so they asked for her arrest. She calmly and quietly spoke with a reluctant deputy at the back of the room before a man more determined to carry out the arrest arrived, handcuffed her, and escorted her out of the room (the video of the arrest is posted on Facebook). The commissioner, in fact, provided the information Alma had requested after she left the room (Ehrlich 2014). Alma's testimony, while incorrect on the policy details of Santa Barbara, was generally correct. The gist of it, that transportation of gas by rail is unsafe and undesirable, is consistent with Santa Barbara County's policies.

Alma spent the next seven days in jail, five of those in solitary confinement without visitors, phone calls, or clean clothes, because she would not give her name, choosing to remain silent (Koch 2015). As the interim director of ACLU Idaho, Leo Morales, said of Alma's arrest: "This is the type of treatment that is usually reserved for terrorists" (Prentice 2014). The jailers claimed they could not process her without a name. She was also jailed before a three-day holiday weekend that inhibited processing. They would not allow her to receive toiletries from her husband, or clean underwear, which have to be approved on Wednesdays—she was arrested on a Thursday (Ehrlich 2014). While in jail, Alma went on a hunger strike. She learned that almost all of the seventeen women in the jail, the "no tell motel" as she likes to call it, were on government assistance. When she discovered that seven of these women with mental health conditions had to pay for their doctors to come into the jail to prescribe their medications, she organized the women to make complaints to the jail. "The next day, they let me out. They probably didn't want me organizing. I'm glad I went to jail since I didn't know anything about it and the mental health situation of the women in there," Alma reflected as she told me about the ordeal over lunch at The Hideaway Grille—the first place she went upon her release from jail four months before. Payette County dropped all charges against Alma of resisting, obstructing, and criminal trespass on April 2, 2015.

One of the best parts about Alma's arrest, according to a couple of interviewees, was that it sparked Jim's involvement (see figure 4). "I will always remember that day, the day Alma got arrested. I had to handle everything while she was in jail, and it was then that I really realized

FIGURE 4. Alma and Jim document new activities along gas lines. Payette County, Idaho, March 2015. Photo by the author.

how crazy it all was. Before, I was aware, but I mostly stayed out of it and did our business," Jim explained. Her arrest made it painfully clear that the way things are handled in Payette County is worthy not only of suspicion, but also of a sustained and huge commitment of his time. He began attending and documenting, with professional video equipment, every meeting he could, and then uploading these to YouTube.

Jim eventually helped form CAIA, where he served as board member and president. His duties in these roles involved organizing meeting agendas and facilitating meetings. When I asked how he described himself, he replied, "Concerned citizen. I believe that there shouldn't be certain elements of society taking advantage of other people and benefit[ting] from them without [it] being a win-win situation." Jim's motivation, like other CAIA members, stemmed from a deep sense of right, wrong, and fairness. He, along with Alma, owned a couple of small businesses (restoration and equipment attachment sales) and applied his business sense and ethics to all things oil and gas. As he said, everything should be win-win. In business, you should provide a quality service to someone and receive a fair price for that service. The oil and gas industry did not abide by this ethos.

This sense of wrongdoing was an important motivator for many interviewees. The behavior of the oil and gas industry sharply contrasted with how interviewees were used to interacting with other people in their community, a community where you often see folks you know at the grocery store. As Luke Smith explained, "I see that the community, a lot of them, just don't know, they are being railroaded and I'm not a protector per se, but I hate to see people [. . .] just getting schemed and railroaded all of the time from the oil company [. . .] and so I've been trying to be educated on the system." Men like Luke, alongside women, felt a sense of duty to protect their neighbors, something that researchers have documented in other rural contexts (see Bell 2013).

Sherry Gordon, self-described "harmonizer," secretary, and website curator of CAIA, whom fellow activist Dottie Hawthorne described as a "quiet force," described a similar realization upon attending her first oil and gas meeting at the Gem County courthouse in Emmett, the town where she lives. A major contention in the fight over oil and gas in Idaho is whether it is about fracking. There are fracking regulations in Idaho policy that the industry participated in developing. In activist's lines of reasoning, why would a gas company work to develop policies for a form of extraction it does not intend to practice? In Sherry's observations, the gas company representatives never say explicitly they will not frack, likely so that if they want to in the future, past quotes will not damage their credibility. However, the company does strive to make fracking seem unlikely. Sherry's concern is that people trust the industry's words:

> The lawyers there were so slick, and it was clear that they were way bamboozling the people in the audience because they knew the right words to say to make implications, but not to say something outright, like, "We are not going to be fracking." [Instead, they say], "We see no reason why we should have to frack here because X, Y, Z." They didn't say they're not going to. And everybody says, "Well, they say they're not going to frack." But oh my God! People are kind of horrifying [light laugh] sometimes in their—and it's not a level of intelligence, it's just wanting to believe something and taking words and not using their best judgment, not using, you know, discretion, logic [laughs].

Sherry's point about the gullibility of Idahoans was something I heard often. Interviewees described Idahoans as particularly trusting, especially of local government. Luke and Brynna felt like the gas company had taken away their ability to trust.[1] Sherry, who moved to Idaho from California, was so concerned about the bamboozling that she agreed to edit an oil and gas ordinance to propose to the county. She did this despite having set out to not get involved with oil and gas because her time was already filled with volunteer work for a number of other community organizations. The oil and gas ordinance absorbed her time throughout the whole holiday season. When she opened their presentation to the committee, she had never done anything like it before, speaking in front of a packed room to recommend policy. Summing up her primary motivation for being involved, Sherry explained, "It's got to be done. [. . .] I mean I really, really, really wish somebody else were doing it, but since they aren't, I just feel [. . .] like I am being held to this somehow. [. . .] I'm doing it out of, oh, you know, a sense of duty I guess, because it really has to be done, somebody's got to be this balance and try to be a force for changing things and I know that those forces all start out small." Her feeling of fulfilling a duty resonated with Jim's account. For Jim, the oil and gas industry was an affront to all things ethical. He feels that ethics do not have a political stripe or label, they are about humanity and doing good:

> The bottom line is, it's not about how happy you are or [. . .] think you are; [. . .] it's about what kind of impact you can make before you check out, you know. Are you, is there going to be a difference? It's not a political thing, it's not a partisan thing, it's a reason why we're here thing, it's a humanity thing. So it doesn't matter if it's left, right, or whatever. A lot of people will coin activists as people that are very liberal, environmental [. . .] but I'm extremely conservative and it's more of a, it's a people thing, it's really—I can honestly say if anybody is concerned about their fellow man, if they have done anything to help support that, I guess they're an activist. So there you go.

The reluctance to claim being an "activist" is evidence that the organizing context of southwest Idaho was one in which doing activist activities was frowned upon. In some ways, interviewees reinforced this sentiment by continuing to distance themselves from the label. However, in their tireless work to stop natural gas, ensure environmental stewardship, and protect communities from being tricked by corporations, they model the dedication of activists. Some, after roundabout explanations, eventually came to the conclusion that they were, in fact, activists. Understanding this social context helps clarify why CAIA developed particular tactics to pursue its goals. These tactics centered around relationship building to enable talking and working across lines of difference.

"PEOPLE DON'T CARE WHAT YOU KNOW UNTIL THEY KNOW THAT YOU CARE"

Relational organizing is central to CAIA members' *theory of change*. Both of these terms come, not from CAIA, but from youth activists whom I interviewed in Santa Barbara. As youth activists explained, relational organizing is the idea that the best organizing comes from personal relationships of trust between people and that these relationships are best built on a foundation of care for the person and their life, rather than the purpose that person serves in a campaign. A theory of change is a person's view of how change happens. I use these terms to illustrate the progressive nature of CAIA's organizing, in the sense that the group is on the leading edge of inclusive and broad-based organizing—across political ideologies. In the realm of political ideology, CAIA is practicing what the youth in Santa Barbara advocate (see chapter 5).

That all core CAIA members recognize the importance of relational organizing is evident in their dedication to message. Here, message does not mean a superficial strategy to attract people, though it is strategic. Message means a core value of the group and the foundation of why they work as hard as they do. They are all committed to their mission of "representing and educating the public, challenging unjust government and corporate actions, and participating in public processes to promote the preservation of private property rights, public health, safety, and resources" (CAIA Mission Statement, approved 11/22/15). In addition, the group has committed itself to six core values of respect, responsibility, trustworthiness, caring, fairness, and citizenship (CAIA Core Values, approved 11/22/15). These values apply to the group, membership, and community. They even apply to representatives of the fossil fuel industry, whom activists like Alma always address with decency in public. Gas company representatives do not always reciprocate. Alta Mesa Idaho vice president and general counsel John Peiserich, for example, once shoved the camera of a reporter sympathetic to documenting CAIA's struggle, and when asked to stop touching the equipment, said "I'll do whatever I want, fuck you" (Koch 2014). CAIA defines its six core values in the following ways:

1. **Respect:** Civility, Courtesy, Honor, Decency, Dignity, Autonomy, Tolerance, Acceptance

2. **Responsibility:** Accountability, Integrity, Follow-Through, Pursuit of Excellence, Self-Restraint, Personal Growth, Humility, Service, Constructive Optimism

3. **Trustworthiness:** Honesty, Truthfulness, Sincerity, Candor, Loyalty, Accuracy

4. **Caring:** Appreciation of Others, Self after Others, Love for People/ Humanity/Life, Giving without the Expectation of Return

5. **Fairness:** Inclusiveness, Equity, Impartiality, Nonpartisan, Nondiscriminatory

6. **Citizenship:** Aware and Informed, Engaged, Do More Than Your Fair Share, Work for Community Wellbeing, Active Oversight of Elected Leaders to Ensure Genuine Representation

By modeling these values and sharing information one relationship at a time, CAIA hopes to build its power. CAIA members routinely spend hours having coffee with people who express interest in their group, getting to know them, their concerns, and how they would like to be plugged in.

These types of relationships are not only important for growing the movement, but also critical for educating people on the complexities of oil and gas. CAIA core members are very concerned with validity, providing sources for all the statements they make. As leader Shelley Brock explained, "The industry can lie 90 percent of the time, but we have to be 110 percent accurate." This requires a large investment in getting new members up to speed. As Sherry Gordon outlined: "You have to spend lots of time bringing [new people] along to the point where you're not really gritting your teeth, hoping that they are not going to say something really foolish that's going to hurt the organization [. . .] just because it's so critical to be [. . .] Teflon coated so that nobody can [put] a grappling hook into you, you know, Teflon coated with truth that's [. . .] back-upable." While important everywhere because of industry's vast resources to amplify its voice above the grassroots, being able to stand behind data is something Idaho interviewees stressed more than other interviewees.[2] They saw data as critical in a context where there is little support for activities perceived as environmental or conservationist.

Justin Hayes, program director at the Idaho Conservation League (ICL), Idaho's oldest and largest conservation group, affirmed the hostile organizing climate that environmental causes faced in Idaho. About six times in our conversation Justin said, "We have to be very careful. If you go with your *hair on fire*, you are marginalizing yourself and all the people and things you stand for. We have to stay credible" (my emphasis). In this case, credible meant middle of the road. Both Ben Otto, ICL's energy associate, and Justin advocated a strategy of having a seat at policy-making tables by proposing what they saw as *realistic* actions— safety regulations and step-by-step victories, rather than a strategy of saying industry was not welcome in Idaho.

Justin's approach to oil and gas clashes with the climate justice movement's keep-it-in-the-ground campaign (see AmazonWatch.org. et al.

2015a, published during COP21). He explained: "Our goal is not to stop this industry, we are not an organization that has said, 'Hell no! No oil and gas development in Idaho period, over our dead bodies!' Um, that would just, that's not a position that is going to work in Idaho, so if you want to make yourself completely irrelevant to the policy debate of how to regulate the industry, go light your hair on fire and say this industry is not welcome here." Probing whether Justin articulated an organizational view or his own, I asked: "Would you personally like to have no oil and gas in Idaho or are you OK with it?" Justin stood and crossed the room to the light switch. He flicked the lights on and off and said,

> These [the lights] come from natural gas. I have an array of solar panels on my house, but am I off the grid? No. I rode by bike here, but then I'll drive my daughter to sports practice, in my Prius, but it's still. So I think we should regulate this to be as safe as possible. It's kind of like with mining [he does mining work for ICL] when people ask me if we need mines, I say yes, for all of the stuff in my awesome cell phone or our cars. If they ask if I'd rather have a mine in Bolivia or the US, I say the US, because we have way better regulations.

Justin's explanation echoes arguments I had heard in Santa Barbara. During a public hearing on the anti-fracking ballot measure known as Measure P, County Supervisor Lavignino made these remarks to support his no vote on Measure P:

> We basically have a soft ban on fracking [i.e., many regulations] since 2011—I voted for it. So this protects us from fracking. The reality is that I had to park a half-mile away from this place because people [who are at this meeting to protest oil] use cars. I am all for solar and renewables, but it [oil] is not going away in the near future. I think *think globally and act locally* is interesting. The GHGs [greenhouse gas emissions] of not getting it [oil] locally is getting it from Iraq or Venezuela, which means bigger GHGs.

These statements and calls for environmentalists to recognize their own carbon dependencies feed into the fossil fuel industry's ability to "manufacture consent" (Herman and Chomsky [1988] 2002; Lippmann 1922). LeQuesne (2019) calls this "petro-hegemony." The fossil fuel industry uses its control over the state, economy, and culture to make fossil fuels an unquestioned element of life. Alternative bases of energy are, in this context, "hardly imaginable" (Herman and Chomsky [1988] 2002:2). In central Appalachia, Bell (2016) documents how the coal industry enacts petro-hegemony by creating a pro-coal fake grass-roots organization called "Friends of Coal" that sponsors local events, services, and places. As she explains: "Through appearing to sponsor everything and anything, Friends of Coal gives the impression that the coal industry is still acting as the backbone of the state, regardless of whether it provides many jobs or contributes significantly to public services. Thus, these diverse sponsorships serve to perpetuate an ideology of dependency: without the coal industry, West Virginians would not only be without jobs, but they would also be without sporting events, soccer fields, cultural events, and community centers" (Bell 2016:104). These corporate strategies are at work in Idaho and Santa Barbara as well. The website slogan for Idaho Power, the utility provider for southern and eastern Idaho, for instance, is "We are Idaho." The website of Santa Maria Energy, a major oil producer in Santa Barbara County, has a slideshow featuring photos of wine grapes and majestic Santa Maria valley landscapes, making it seem as if the company is somehow synonymous, rather than incompatible with the county's largest economic sectors—tourism and agriculture. Only two of the four photos in the slideshow feature any oil infrastructure. In one of those, the infrastructure is almost completely blocked from view by trees.

Convincing anyone who uses fossil fuels that they are dependent on them, that stopping fossil fuels would lead to the ruin of the individual and society, bolsters the hegemonic power of the industry. It prevents imagination of new and different ways of obtaining energy and organizing society. It also divides community members, marginalizing those that are calling for what we need: no more fossil fuels. That this dependency message comes from the program director of Idaho's largest environmental group and from an elected official in Santa Barbara demonstrates the broad buy-in to industry's messaging. It is as if the American Cancer Society said, to a person who smokes daily, that the inconvenience of quitting and adjusting one's daily routine outweighs the known health effects, and that the smoker should buy local tobacco (see Oreskes and Conway [2010] on the parallels between the tobacco and fossil fuel industries). Grassroots groups like CAIA work hard to negate this fossil-fuel dependency justification for inaction.

Unlike the nonprofit staffers I just mentioned, CAIA thinks the facts about the damage of gas extraction give them the credibility to advocate for at least a reversal of all the laws that make gas extraction a viable business in Idaho, if not a ban on the industry. The facts CAIA shares with communities and their concern with progress on the issue endears them to their supporters. Care, as Jim's quote in this section's title communicates, is key: "People don't care what you know until they know that you care."

Just one example of how CAIA showed people that it cares comes from Peter and Susan Dill, organic farmers in Gem County. They stressed how impressed they were that Alma had been responsible for connecting their county (where Alma does not live) to Michael Lewis, then director of the United States Geological Survey's (USGS) Idaho Water Science Center. They did not know Alma had been involved and had wondered how Lewis connected with Gem County. With guidance from Lewis, the USGS partnered with Gem County to conduct county-wide base-line water testing *before* industry drilling. CAIA members believe this is the first such collaboration in the United States. Having baseline data is the only way communities can prove contamination from oil and gas drilling; in most cases, communities do not have this data. The Dills, both soft spoken and intentional with their words, spoke highly of Alma and Michael. According to Susan, Michael "was very supportive, I mean he really *cares*" (my emphasis).

In the context of climate crisis, care means shutting the oil and gas industry down. Though the Idaho Conservation League's website is full of language about working on the issues Idahoans care about— "Because you love Idaho, the Idaho Conservation League protects the air you breathe, the water you drink and the land you love" (Idaho Conservation League 2015)—their moderate approach to oil and gas regulations fails to care enough. In trying to walk the middle line on this issue, to appear "credible" to Idaho politicians (many of whom do not believe in climate change), they alienate CAIA and fail to demand the conservation measures that climate change requires.[3]

Though taking a moderate approach may appear to be an attempt to talk across lines, it is an attempt to talk across, but not disrupt, *lines of power* in a political economy that is fundamentally unjust. Both Ben Otto and Justin Hayes were reticent to think about building coalitions with CAIA, whom the former saw as representing conservative voices that had been against most of ICL's policies in the past, and whom the latter saw as people with their hair on fire. Both had had little sustained personal contact with CAIA members and no relationship of trust. Ben Otto recognized the importance of trust. When I asked him if there was a way ICL could collaborate with CAIA on oil and gas, Ben said:

> We are definitely trying to. [. . .] It's new for us, we have traditionally just been very threatened by the Tea Party and private property rights scene and they have felt very threatened by us, so it's new for both of us, both groups, to try to find this comfort and trust. It takes a lot of trust building when you've traditionally, you know, lobbed competing press releases at each

other and you've called each other terrible names in meetings; there's a lot of repair that needs to happen, and trust is earned, especially in Idaho. I mean people are pretty insular to their community, so it is just going to take time to build that trust and you earn it by demonstrating consistency and respect for different people and listening to folks, but I think it's getting better, but yeah, it will take time.

Both Justin and Ben recognized, in their own ways, where CAIA was coming from. Justin noted how it was understandable that people radicalized. He had seen activists start off concerned and "reasonable," but then, when no one listens, he explained, people have no incentive to be moderate. Ben, who was perhaps not yet so jaded to dismiss what he called "the left flank" and who personally preferred that oil and gas did not happen, thought it was important for people to make more radical demands than he could in the context of his more insider position with ICL. Radical demands like "Let's close every coal plant this year" moved the window of possible conversations to the left, Ben explained. Ben and Justin felt they were doing what they could within their roles as ICL staff—lobbying for more stringent policy and building strong relationships with state agencies. During my research, their practice of doing what Ben called the "art of the possible" did not line up with CAIA's goals.

In this context, CAIA, particularly Alma, perceived lack of support from the big green groups in Boise, and so, most of CAIA's relational organizing was focused on building the grassroots. Of CAIA interviewees, Jim thought about relational organizing the most. He was a keen observer of body language to gauge a person's feelings. As he explained, activists should be focused on others' needs and be aware of how different issues concern different people. In the following excerpt (and table 1), he powerfully captures the idea of talking across lines—the core of CAIA's organizing:

> You've got to figure out a way to establish that you care, and to do that you have to [. . .] understand where they're coming from and learn how to speak to them. Instead of telling them everything that *you* think they should know, you need to tell them things that they would be interested in knowing. [. . .] If you can imagine taking a piece of paper and folding it in half, [there's] the left side and the right side and the line in the middle is the center; list all the things that different political affiliations would land on. So, obviously on the left, you would have abortion, and abortion on the right too, you would have maybe guns on the right and you know, list everything, and then try to find things down the center that people, that both people, both mindsets, would see [as] a common interest. [. . .] And, you know, see what goes down

TABLE 1   THE PRACTICE OF TALKING ACROSS LINES

Left	Center	Right
climate change	community	climate change
environment	health	small government
abortion	home and property	abortion
guns	water, air, soil	guns
	government accountability	
	business integrity	
	public infrastructure	

the center and then that's what you talk about. If you're in an area that is more to the left, then you can put that line over to the left, or more to the right, but you have to kind of look at first an overall view of what you want to do and you try to stop talking about the fringe stuff because that's what they [the oil and gas industry] want us to do.

I depict Jim's idea in table 1, illustrating how an activist's task is to understand, through relationship building, where someone is coming from. You meet someone where they're at, as Santa Barbara youth activists say in chapter 5—and then develop a strategy for communicating with them.

Jim also used the example of farmers, an important constituency in a strong farming community like Payette County, to explain what talking across lines would look like in his organizing context:

What is important to the farmers? Well, water is very important, make sure they got plenty of water for their crops, because I don't think there's a farmer around that would say, "Oh no, I don't care if water gets cut off midseason." I mean that's their source of life—they don't bring that product to market, they don't get paid. They got a lot of money up front, they may have even taken loans to be able to put the product in the ground [. . .] so you have to look and see what's important to them and then actually write a list, so you can keep that top of mind, so water would be important to farmers. Condition of roads are important; they can't get their product out if the roads are not good for large trucks.

The things that Jim identifies, water and road conditions, are both directly impacted by natural gas production. Hydraulic fracturing uses tremendous quantities of water and the heavy trucks that carry water and natural gas infrastructure.

Jim's analogy, however, goes beyond oil and gas. His approach underlines common values as key to organizing. While there is some flexibility in communicating values depending on context, the core values that drive CAIA's work are always the same, rooted in care, fairness, and quality of life. The goal of protecting these values leads clearly to some common enemies, in this case, the oil and gas industry and the state. By standing on the core values at the center of Jim's imaginary piece of paper, CAIA can talk and work across lines of difference that typically prevent collaboration.

CONCLUSION

Through an exploration of activists' journeys into activism, this chapter describes the social context in which activists work. It is a social context where activists are seen as extreme and unreasonable and where activists perpetuate this image, at least for a time, by distancing themselves from the term. It is a context where leaders in the environmental nonprofit realm perpetuate the idea of fossil fuel dependency, that there is no alternative. Despite this context, most interviewees eventually recognized their work as activism and as critical to securing a healthy environment and fair political context for their communities. The natural gas companies' violation of rules of decency for relationships was a spark that lit the passions of most interviewees. On the flip side, activists' relationships with each other helped them overcome feelings of isolation that result from organizing in such a social context. Similarly, CAIA was able to rapidly grow through emphasizing values like care, understanding the priorities of community members within its messaging, and then talking across lines to highlight common values.

In chapter 4, I describe how Idaho activists talk across lines, as well as the challenges to doing so.

CHAPTER 4

Idaho Part 2

Talking across Political Lines
by Agreeing to Disagree

In southwest Idaho, activists had a particular approach to working across lines that I call *talking across lines*. As I describe in the preceding chapter, *talking across lines* is about knowing your audience and finding things down the center, common values that resonate with you and your audience and are jeopardized by oil and gas. The dedication to drawing on common values to communicate about fracking, connect and build relationships with community members, and mobilize and empower new activists arose repeatedly in my time with CAIA's members. Alongside water and road conditions (core concerns of farmers), another theme that CAIA emphasized to convince homeowners of the importance of resisting natural gas was private property rights. These concerns, voiced in terms of care, fairness, and quality of life, enabled CAIA members to talk and work across lines of difference that typically prevent collaboration. In this chapter, I analyze how people succeed and fail in their efforts to talk across lines, using the fight against natural gas development in Idaho as a case.

Talking across lines works when activists put aside political affiliation and beliefs about climate change, expose the *roots* of injustice, and focus on core values of respect, responsibility, trustworthiness, caring, fairness, and citizenship (see CAIA's definitions of these in chapter 3) as motivations. CAIA articulates the roots of injustice as lack of integrity and accountability. When people can come to the organizing table around shared values, they can build relationships and trust, and then, perhaps, move on to changing their ideas, values, and goals as they learn from each other. Mobilization is about movement of the heart (Tsing 2004); core values pull at the heart, calling a person to action. In effect, talking across lines builds unity against the fossil fuel industry's efforts to divide.

However, this practice is not neat or always successful. Like any social process with progressive potential, it is rich with friction, to borrow from Anna Tsing (2004). Talking across lines requires an openness to messy, incongruous, and shifting perspectives. To illuminate the complicated nature of talking across lines, I explore how this messiness manifests within individual beliefs and social interactions that change over time.

Talking across political lines happens when people with differences that typically divide are able to come together for collaboration on a substantial common goal. I say substantial, rather than long-term, because CAIA had only existed for about two years when I conducted my research. They were, however, working on a substantial goal in that their lawsuits required large time commitments to continue raising money for legal fees. Filing a legal suit together is a substantial endeavor and works toward a substantial goal, in this case, overturning the legality of forced pooling in the state of Idaho.[1]

CAIA relied on identifying shared values and concerns to talk across lines and build unlikely alliances. Defining common values, messages, and goals, as well as those things that are not part of the alliance, required deliberate ongoing work on the part of CAIA's members. In the next section, I describe how CAIA's messaging centered common values.

MESSAGING PROPERTY RIGHTS AND PUBLIC HEALTH

As Jim hinted in the previous chapter, the language used to describe core values is important. CAIA had consensus that property was the most effective way to communicate core values across a broad base of Idahoans.[2] Idaho has a high home ownership rate, 70 percent versus 64 percent nationwide (US Census Bureau 2019a). In all but one of the towns where CAIA was organizing, homeownership was even higher—70, 74, and 80 percent. Agreement on this message, however, did not characterize resistance to oil and gas from the beginning. Alma and Tina Fisher, who were the first to start publicly resisting oil and gas in Idaho in 2010, originally used an environmentalist message. They formed a group called IRAGE, Idaho Residents Against Gas Extraction. They did not have much success garnering local support, but did connect with fractivists in other parts of the country. With sponsorships, both Alma and Tina went to Washington, D.C., in 2012 and Dallas in 2013 for the annual Stop the Frack Attack rally.

While the knowledge and connections Alma and Tina cultivated before 2014 were paying off, locally, Alma thought, "I would've been more effective probably just beating my head against the wall, quite frankly." Their environmental message—even when communicated in non-buzz-word language, like "Aren't you pissed off about the potential impacts to our aquifer?"—created a "vise grip" on people's heads. In Alma's view, "You got to get the screws unloosened on the vise grip and get [it] off people's heads [. . .] before they can listen, because the reality is, once they understand the corporate dominance behind what's happening and how we got from point A to point Z, they'll get all kinds of pissed off and you'll have somebody who becomes an activist pretty quickly, who would ordinarily not." On the other hand, Alma explained that environmental messaging turned people against CAIA's cause: "If you start [with], 'Aren't you pissed off about the potential impacts to, you know, our aquifer?' you know, there's a good chance you might be labeled, you know, one of those *damn environmentalists*."

After experiences like this and suggestions from Jim and Joe Morton (who are both more conservative than Alma) to change the way she communicated, Alma began to focus on how gas threatened property rights. It seemed to work. As Alma explained,

> If you start that same conversation saying, "Geez, George, you know that they're going to be able to force you into doing this [lease to the gas company] and if they force you into doing this and something happens to your well water, what are you going to do? You know, then your property's worth nothin." You know, that gets them thinking, and it—and once you get people thinking, you know, if, once the barrier is up, there is no conversation. So you've got to, you have to figure out a way to get a message across without enacting those barriers, and I think we've [CAIA] been pretty good about doing that.

A core piece of the property rights message is the effect an oil and gas lease can have on a mortgage. Alma, who worked in the mortgage industry for fourteen years, had a letter from her mortgage lender stating she would be in technical default of her mortgage if she signed an oil and gas lease. From the mortgage lender, or insurance company's perspective, an oil and gas lease can damage the property; some lenders will only make a loan to people who agree not to lease (Urbina 2011). Residents of Oklahoma, a state that has now experienced more earthquakes than anywhere else in the world because of fracking waste water injection wells (Chow 2015), have been struggling to find insurance coverage for their frackquake damaged homes (Summars 2015). Soon after Alma switched her environmental message to property rights, support grew.

An additional strength to this message is that despite the distaste for environmental messaging that conservatives like Jan and Wayne, former leaders of the Tea Party group in their county, express, environment is embedded in notions of property. Property gets people thinking about their land, air, and water—in other words, the environment.

Corrie: What do you think is the strongest message to get people on your side around here?

Jan and Wayne: Property rights.

Wayne: You have one shot to get it right, you mess up and they are going to ruin your aquifer. One spill on the soil is going to destroy that soil; you cannot regenerate it once you've had that spill. [. . .] That soil's gone; you don't have a second chance.

Here, Wayne, who is adamant about the need to be nonpartisan in all things oil and gas, which includes avoiding the word *environmental* because it is too associated with liberals, moves seamlessly in our conversation from underlining the importance of the property message to talking about protecting water and soil. The property message can extend to public property, the commons, as well. CAIA's mission statement makes this extension: "private property rights, public health, safety, and resources." Public resources implicitly include air, water, and land.

The most important part about the property message is that it opens the door to conversations among people with different perspectives, yet who can identify with the importance of protecting their homes and/or property. The property message permits a conversation to develop without barriers popping up or vise grips on heads tightening in response to topics that are polarizing in the local context, topics such as environmentalism and climate change. In sparking dialogue, the message of property rights lets relational organizing and relationship building begin by allowing people with divergent opinions to get to know each other as they work against a common enemy and discover their shared values—quality of life and accountability from government and industry. By fostering trust, the relationships built around shared values make the group more effective in their work. Eventually, trust also allows people to have hard conversations about their disagreements and, potentially, change their opinions.

There are of course downsides to the property message. Most importantly, there is a risk that someone will be concerned *only* about their property, never moving past the NIMBY (not in my backyard) or NUMBY (not under my backyard) stage.[3] However, among interviewees, it was more common to progress out of NIMBYism toward a recognition and concern with the broader impacts (our backyards) of natural gas, or to express concerns that blur and complicate NIMBYism. As Brynna Smith explained, "Initially, I would have been like just put it [the gas well] somewhere else, don't put it here, I don't want to deal with it, but now I think, 'Oh my gosh,' you know, they put it somewhere else and then some other family's in the same position that we are in." I discuss these complexities in depth in the following section. Suffice to say that personal experience, damage, or threats are strong motivators and motivation is the first step in long-term organizing.

Sarah Pierce, for example, became a CAIA member after landmen trying to get her to sign an oil and gas lease visited her home three times. The third time, three men came, barring her from closing the door when she repeated her refusal to sign the lease. Sarah, alone with her children, "felt very intimidated by them." They told her she would be an idiot not to sign, but she refused and successfully demanded they leave. Joli, who owns a dog training facility with her husband, got involved when Payette County approved Alta Mesa to build an ancillary processing facility and rail spur on the lot adjacent to their property. There would be a flare stack six hundred feet away from her home and dog kennels and a constant hum. Both would make her business impossible because of the effect on the dogs her husband trains, who would have constant exposure to the chemicals and noise. As Joli explains, Planning and Zoning, who had approved an expansion of Joli's business (a large investment for her and her husband) just about a year before the Alta Mesa project was proposed, "with their eyes wide open, sold us down the river." "Everybody in the entire neighborhood that had anything to say about the plant said no, and they passed it anyway," she said.

Examples like these and the Smiths' story described previously illustrate industry behaviors that interviewees interpret as unfair and unjust. These stories are powerful in a context where drilling has yet to produce any catastrophes that would make the environmental threats seem real. With only eighteen wells permitted by the end of my research, fifteen of which were drilled and nine of which were producing, it is not surprising that there had not been any publicized accidents or violations. Thus, at this phase in natural gas development, people's stories of threats to their families, homes, and way of life were more powerful organizing tools than statistics on environmental impacts. What the state of Idaho and industry can do to an individual's property, without their consent, makes real the threat of natural gas for anyone who owns a home or knows someone with a home near oil and gas leases. The message sinks in rapidly, avoiding much of the difficulty that arises when trying to communicate the process of fracking and its effects. Alma for example, sometimes refers to industry's definition of fracking: "high-volume, slick-water, horizontally drilled hydraulic fracturing"—quite a mouthful. The topic of property rights, when combined with all of Idaho's oil and gas policies, powerfully illustrates industry's and the state's disregard for CAIA's sense of right and wrong. Shared understanding of the injustice at the root of oil and gas, an injustice that extends beyond this industry to others (e.g., industrial agriculture), strengthens relationships among CAIA members. In Bhadra's (2013) terms, "disaster scripting" about oil and gas reworks social divisions "into a common identity of shared imagination of disaster." Injustice was manifested not only in CAIA member's interactions with the gas company, but also in state policies that favored industry over community, which I now discuss.

"EVERYTHING WAS HANDED TO THEM ON A SILVER PLATTER"

Joli's case in the previous section is a microcosm of how government collaborates with industry to approve plans that disrupt people's lives. There is agreement, as Wayne says, that the oil and gas industry has things "handed to them on a silver platter." As Peter Dill highlighted, industry is not a good neighbor. Along with other interviewees, Peter had seen the gas company's representatives speak at meetings. He was not impressed:

> They've got this guy [Michael Christensen, a lawyer who represents Alta Mesa] who comes to the meetings and says, "Well, you know we should be treated like all the other neighbors, how come we're not?" And he seems to really wonder that and everybody else says, "Well, there's a reasonable reason: You're not a neighbor; you're representing a corporation, which by the way, we gather is on its last legs financially and trying to bail itself out. But there will be more behind you; you're not our neighbor. When we [. . .] realize that you are going to come and go after kind of leaving a lot of damage behind, we're not exactly friendly." I think people get stirred up by that, by someone doing damage and not paying for it and leaving the bill with us.

With reference to Peter's comment that the corporation "is on its last legs financially": Moody's Investors Services, a credit-rating firm, downgraded Alta Mesa's financial status in 2014 from "stable" to "negative" (Smith 2014b).

At the state level, Idaho has a practice of adopting policy put forward by the oil and gas industry and other industry-friendly states. For example, in 2011, Idaho adopted rules based on oil and gas procedures in Wyoming; a few months later, a report by the US House Energy and Commerce Committee revealed oil and gas companies operating in Wyoming had injected thousands of gallons of carcinogen-laced water into wells from 2005 to 2009 (Prentice 2011a). As a representative of the State of Idaho told Luke Smith when Luke inquired about setbacks: "Well in Texas they can put it right next to your house; they don't mind." In 2015, Idaho had no setbacks requirements for schools, playgrounds, parks, hospitals, or residential areas (IDL 2015), advocating that "spacing of wells must achieve the goals stated in Idaho Code § 47-315, one of which is to maximize the recovery of oil and gas" (IDL 2015:11). This essentially requires maximizing profits. In 2016, the minimum setback was set at three hundred feet (Barker 2016).[4]

"Really?" said Luke, taken aback by the idea that putting a well next to your house would be alright. The representative said, "Yeah, if it's good enough for Texas it should be good enough for here." As Luke explained, "I'm like 'What?! [laughs] and all the earthquakes in Texas don't alarm you at all?'" There are countless similar instances, where elected officials in Idaho downplay risks and promote industry-friendly regulations. CAIA works hard to expose this reality to the public.

The two recently passed bills that CAIA frequently highlighted during my fieldwork are House Bill 50 and 464. House Bill 464, passed in 2012, strips local control over oil and gas, forbidding local governments from enacting local bans. Cities and counties cannot require oil and gas exploration companies to secure conditional use permits for their projects (Associated Press 2012). Joli, on the other hand, who owned a dog training business with her husband, had to get a conditional use business permit for their small business! One of this bill's supporters in the legislature, Monty Pierce, leased his land to Snake River Oil and Gas, the company behind the bill. He waited for the final vote to disclose his leases and then, when Democrats filed a conflict-of-interest complaint, the Senate Ethics Committee dismissed the complaint (Associated Press 2012). The irony of this bill, completely incongruous with Idaho's reputation for hands-off government and local control, is thick. By educating the public on the bill, CAIA illustrates the hypocrisy of Idaho representatives and also puts Idaho's plight in the context of the larger anti-fracking movement in the United States, where states across the country have stripped local governments of control over the oil and gas industry (Healy 2015).

As in the rest of the United States, the oil and gas industry also backs politicians and regulators (Turnbull 2016). Bridge Resources, the sole natural gas developer in Idaho until it went bankrupt in 2011, contributed money to each of the five members of the governor-appointed Idaho Oil and Gas Conservation Commission, with a five-thousand-dollar contribution to Governor Otter (Prentice 2011b). This *appointed* commission was given regulatory power over oil and gas in 2013 through Senate Bill 1049, which stripped this power from the body formerly charged with regulating the industry—the *elected* State Board of Land Commissioners. Cases like these bolster CAIA's message of government corruption.

House Bill 50, passed in 2015, details the process by which industry can ask the state to *force pool*—"integrate," in industry terms—mineral rights owners in sections of land. A section of land is 640 acres. Under House Bill 50, once the gas company gets owners of 55 percent of mineral rights in a section to agree to lease, the owners of the other 45 percent can be forced into leasing.

Forced pooling gives immense power to large landowners, many of whom wield great political power. For example, Brad Little, Idaho's lieutenant governor from 2009 through 2018, and governor since 2019, is a mineral rights owner in and around areas that have been leased for oil and gas development near Boise (Malloy 2015). The Idaho-based Simplot family, who pioneered frozen french fries, supplies McDonalds, and promotes genetically modified potatoes, was the seventeenth-largest landholder in the United States during my research (*Land Report* 2015).[5] Butch Otter, also a large landowner and Idaho's governor from 2007 through 2018, is the former son-in-law of J. R. Simplot (Yardley 2010). The Simplots are not only large landholders and intertwined with the political class in Idaho, but also fund pro-industry politicians. The Simplot political action committee has overwhelmingly supported Republicans over the years, including Idaho's congress members Mike Simpson, Raul Labrador, and Mike Crapo (Center for Responsive Politics). All three politicians support America's energy independence, "accomplished through a combination of renewable energy, nuclear energy, clean coal, and both onshore and offshore oil and gas" (Simpson

2010).[6] In other words, they have an all-of-the-above energy policy that ignores climate change. Simplot likely funds pro-industry politicians to decrease the very environmental regulations that it has violated. In 2015, Simplot paid $899,000 in a civil penalty to settle a number of Clean Air Act violations and agreed to spend $42 million on improved emissions controls and monitoring (US EPA 2015). Large landowners in Idaho like Simplot, Otter, and Little, who stand to gain the most from oil and gas development, are also intimately tied to the political and economic power of the state. They are generally pro-industry and against environmental regulation.

Idaho's lax oil and gas policies and their entanglement with the interests of elites add fuel to CAIA's fire. CAIA marshals examples like these to advance its central message: the state's support of oil and gas is an egregious misappropriation of state resources that should be directed to representing and pursuing policy in the best interest of the people. By framing the oil and gas industry as an industry receiving unfair benefits, CAIA advances a message that is broadly appealing to conservatives and liberals and especially to small business owners in the area where CAIA organizes. CAIA diagnoses the problem as state and industry collusion, offering the solution as integrity and accountability. As the header of their website read, "Citizens Allied for Integrity and Accountability THE PEOPLE INSIST" (CAIA 2016). Though exposing injustice in the three Ps of property, profits, and politics brings CAIA members together, it does not wash away all differences. "The People" are not monolithic. Talking across lines also requires acknowledgment of differences of perspective. As the most challenging part of the practice of talking across lines, acknowledging difference can determine collaborations.

"CHECK YOUR PARTY AFFILIATION AT THE DOOR": ADDRESSING DIFFERENCE IN CAIA

CAIA approaches difference in two ways: by checking it at the door and agreeing to disagree. On one hand, these practices ignore difference. On the other hand, by being so explicit about how to maneuver around and work through difference, they embrace it. Before I describe these practices, a word on defining difference. Here, difference refers to difference of perspective or belief—political difference. Since CAIA is a relatively homogenous group in terms of race, ethnicity, and sexuality, though its membership does reflect class diversity, CAIA's practices around different *perspectives* may not extend to other forms of difference. One core CAIA member identifies as a Native American, but not as a person of color. This person is the only core member without primarily white/European heritage. At CAIA events, I have hardly ever seen someone I perceived as a person of color. In a state where 93 percent of the population identifies as solely white (US Census Bureau 2019a)—a proportion that mirrors most of the towns where CAIA organizes—this is not uncommon.[7]

All CAIA core members are in heterosexual marriages. To her credit, Alma is an outspoken supporter of gay rights, a prominent civil rights issue in Idaho because of the state's continued support for legal discrimination against LGBTQIA+ individuals, and she has built relationships with Duck Valley tribal members. On the other hand, some group members' views align with conservative views on abortion, gay marriage, and immigration, and, though not evident in personal interactions, the fact that some of these perspectives are public on Facebook could prevent people from particular marginalized communities from wanting to participate in the group.

My own politics contrast sharply with those of some group members. Despite the impression that some group members' views exude, my interactions with these same members demonstrated, over and over again, that they are well-meaning individuals interested in the welfare of people and communities. In line with the thesis of this book, that relationships of trust are necessary for building broad-based social movements, I see intolerant views stemming from an absence of relationships of trust with different people. Building trust between people supporting policies detrimental to marginalized communities and members of those marginalized communities is obviously difficult and an issue of safety. I contend, nonetheless, that with a commonly shared value and an effort to build a relationship working around that value, people with disparate experiences and beliefs can begin to share conversations, learn, and ultimately, persuade and change each other's ideas in the interest of justice. Justice is itself a core value whose basic definition I and interviewees likely understand in the same way, but that fact is often obscured by media and corporate efforts to instill different understandings of policies through which political parties try to put justice into practice.

While recognizing the serious problems that individual public displays of anti-immigrant sentiment, for example, pose for inclusive organizing, CAIA's *group* strategy and mission *is* beneficial for creating *political* inclusivity and holds potential for fostering broader inclusivity. Thus, while my argument in this chapter addresses possibilities of talking and working across *political lines*, I see evidence from my other research sites that CAIA's practices can facilitate collaboration across other lines as well.

Checking partisanship and views about climate change at the door was a cornerstone of CAIA's approach to organizing. As Wayne's quote in this section's title—"check your party affiliation at the door"—communicates, core members chose to meet around Dottie's table as people working on a common goal. They eschewed political labels, recognizing their use as a dividing tactic by the oil and gas industry. Though they held various views that are usually understood as belonging to one party (e.g., pro-choice, pro-life, suspicion of taxation and welfare, support for gay marriage) they all agreed that "neither party is worth a hoot," as Dottie Hawthorne put it. Many had been registered Democrats and Republicans at different points in their lives. During my fieldwork, however, most core members thought of themselves as Independents, conservatives, or liberals, and in some cases, they tried to avoid all such labels for their beliefs.

While they tried to explicitly check their political views at the door, they were aware of each other's views. With their awareness, they chose to agree to disagree. Dottie, who had decided to identify herself as an Independent rather than Republican, explained the process:

> You just have to allow everyone to have their own opinion and don't degrade them for it because they have their reasons for believing how they believe and yeah, oh, my political differences are way out the window, I mean really, really different, and I have to marvel because I just love them all [CAIA core members], but some of them who are oh, very, very intelligent, have the weirdest ideas about things. I think, "What, as smart as you are, you actually believe that?" you know. But that's just me. [. . .] I hope that one of these days that they will say, "Hmm, maybe I wasn't right about that; maybe I should rethink this" [chuckles]. And maybe they think the same thing about me. I don't know, but I try not to say either way. I just keep my opinions to myself about, especially political, because, that is ingrained in most people from youth, actually, and being a rancher, you know, I am Republican.

That Dottie calls herself a Republican at the end of the quote, affiliating herself with what the party used to be despite her current aversion to the label, illustrates the complexity of Dottie's relationship with political labels. For Jan, Wayne, and her, Republican was no longer a good descriptor of their beliefs because they saw the party as a RINO, Republican in Name Only—insufficiently conservative. Wayne, for example, valued small government. He saw local Republican politicians' support for oil and gas development, and for the erosion of private property rights that that development entails, as government overreach.

While core members had decided to keep their differing views to themselves, they did not always make this practice clear to new members. I had not seen a CAIA member explicitly explain the importance of nonpartisanship to newcomers at CAIA events or meetings. I learned how important nonpartisanship was to CAIA's identity mostly through my sustained one-on-one interactions with Alma. When Tom Cervino started coming to CAIA meetings during my fieldwork, Dottie unofficially took on the task of maintaining the group's nonpartisanship by making reminders during meetings. At a CAIA meeting on October 14, 2015, Tom mentioned that he knew a couple of people active in the Republican Party who were against fracking. He wondered, however, if they would stand up or not. Would they support CAIA and make their opposition public? Tom's comment implied that Republicans would be less likely than Democrats to stand up against fracking. In reply, Dottie said, "I don't think it is a Democrat or Republican issue." To her, it was important for the group to talk about politicians and parties as a whole, recognizing that both major parties have poor oil and gas policies. While it was rare for CAIA core members to talk explicitly about their disagreements, conversations were possible. The most powerful example of such a conversation is my interview with Alma and Jim.

Alma believed in climate change; her husband Jim was a climate skeptic, so they practiced talking across lines in their relationship as spouses. Though they did not talk about it much, they did discuss it in our interview.

Corrie: Can you tell me about your perspectives on climate change and how it is to have your spouse have a different view?

Alma: [laughs] Oh, now you are lighting a torch! I love it.

Jim: Um, that's one of those things that I think in CAIA we agreed not to discuss because especially here in Idaho, you can really alienate people, and that's getting that piece of paper out and putting things down on the right side, and the left side, and the center and trying to get as much stuff to the center and making that your topic. Um, again unless you're working in different areas where you can adjust that centerline.

Jim's main objection to acknowledging climate change was that politicians and corporations control the media; therefore, he believed that addressing climate change is another example of opportunistic unethical business. Here, he referred to Al Gore and the money Gore has made on carbon credits (see Broder 2009). Ironically, in this regard, Jim is in line with climate justice activists who critique market-oriented approaches to addressing climate change as false solutions (see Cabello and Gilbertson 2012 and Reyes 2012). Alma, in contrast, described herself as a "firm believer in climate change," explaining,

> I think the science and the data are very clear, and obviously my husband and I are at total polar opposites ends of this discussion. I hope that I'm wrong, I hope that the scientists are wrong, I hope that the data is wrong and [. . .] I will happily go and eat my words till the cows come home. I hope and pray that will be the case. I am terrified it is not.
>
> And I think every day, we are December 12, probably going to be a fifty-degree day in Idaho, we've been getting rain and not snow and we've had four storms, four thousand-year storms in a six-week period of time, I mean it's just, I think that the data is clear.
>
> That said, because we're in Idaho, I don't have climate change conversations with people. I do put up climate change information on my personal Facebook page [. . .] hoping that this will prompt them [her Facebook friends] to read things and look at the data and that kind of thing; it may or may not. But as far as CAIA is concerned, we don't, you know, take an issue on climate change. That said, I personally know that if we are successful in stopping oil and gas activity here in Idaho and possibly for a precedent-setting lawsuit across the country, that is probably one of the most single, largest things that we can do to have a positive impact to actually start to reverse climate change. [. . .] I'm OK with actually not engaging in it in a vocal manner in Idaho because I feel like the potential for what we can accomplish by actually not being vocal about it, by working on the oil and gas issue and getting people from across, from *both* sides of the aisle to work with us hand-in-hand on putting a stop to it, has way more significance and importance than the actual conversation here in Idaho. Because it's not, we're not going to get anywhere in the conversation because there are a lot of people who feel like Jim. [. . .] So anyway, that's just, we have, that's just one of those things in our home that we can agree to disagree.

In reply, I asked, "How did you learn how to agree to disagree? It is a very important skill that most people don't have, it seems. What taught you how to do that?" "Survival," Jim replied. Alma continued,

> [laughs] It's one of those things that [. . .] it's sort of like he's entrenched in his position and I'm entrenched in my position. [. . .] I think ultimately what will bite him is when it becomes very obvious and clear that it's here, or, I will budge because it becomes very obvious and clear that it's not. [. . .] Ultimately, you know, worst-case scenario, we work to improve [. . .] our air quality, we work to improve the quality of our water, we work to improve and transition from fossil fuels where corporations control us, to alternative and renewable energies where we control our own destinies, you know. And

I think we could do a much better job as a community and as a movement messaging that way. [. . .]

That's actually a message that I've had some degree of success in communicating here in Idaho to climate change skeptics, *not* bringing in the climate change conversation but just saying to them, "Well, you know what, so long as they can control what happens when you flip that switch, you don't control your own destiny, somebody is always in power, has power over you and control over you. Only when you control what happens when you flip that switch, or you turn that ignition on in your car, that's freedom, that's ultimate freedom." *That* people get; that they connect to, but if I go and say to them, you know, because of climate change you better get some damn solar panels on your roof, you better get that electric car in your garage, that will not go over well. [. . .] So, yep, we just agree to disagree; you have to, you know.

Jim: Plus, we don't even talk about it, there are so many other things. Not a big concern. We've never directly even talked about it.

Alma: We have, I've tried to send you things.

Jim: Oh, you tried, I said we—two-part communication [all laugh].

Alma: And he won't and that's OK.

Whether or not there is a deeper tension behind Jim's humor in this exchange is not clear. Alma and Jim had been married for decades and worked closely with one another in business and CAIA. In the seven weeks I lived with Alma and Jim, our interview was the time I saw them most at odds. Their ability to agree to disagree with one another amid the stress of organizing 24/7 demonstrates their skill at the practice and bodes well for CAIA's ability to continue building a group across political lines.

However, if not grounded in a strong relationship of trust, conversations like Jim and Alma's can endanger members' working relationships. CAIA lost two core members over a heated conversation related to partisanship where members with divergent views were unable to agree to disagree. The conversation happened on the heels of public forums that CAIA hosted where, in addition to encountering technological difficulties, organizers disagreed on how or whether to communicate about solar energy to conservative crowds coming to learn about oil and gas. This phase in the group's development demonstrates that talking across lines is not easy and not always neat or successful. Besides commitment, its success is dependent on individual personalities and stress levels as well.

Talking across lines begins when an issue of common concern brings people together. For CAIA, this was accountability and integrity, messaged through property. Then, the violation of common values by a common enemy strengthens desires to take action. This is where Idaho's policies and elite interests illustrate the injustice of oil and gas development and the need to change state and industry behavior. This exposure of injustice is an ongoing process of talking across lines. To continue working together, members must cultivate the skill of agreeing to disagree. With enough trust, conversations about these disagreements can take place, and even convince people to think more about each other's perspectives, as demonstrated by Dottie's and my exchange in the previous chapter. Jim, also, over the time I conducted fieldwork, became less and less adamant in his climate change skepticism. The components of talking across lines I have identified build unity against corporate efforts to divide, something of vital importance for all movements for radical social change.

CAIA's work to talk across lines provides insights and practices that may work particularly well in places with diverse political stripes. Understanding how to talk across political lines, however, also requires understanding the messy nature of difference itself. In the next section, I describe how people resisting natural gas in Idaho disrupt the stereotypes that are associated with certain political labels. Jan's caution to not bring up politics because of the preconceived notions people will have about who is for or against fossil fuels is appropriate. However, if people were to talk openly about politics, they may discover that folks with different ideologies are not as far apart on views on oil and gas as they may assume. In the cases I describe in the pages that follow, political party affiliation is not a very good predictor of people's beliefs.

FRICTION: THE MESSINESS OF TALKING ACROSS LINES

How do we convince people that life on earth is worth saving? This
is a political question as well as an environmental one. It requires a
politics of working across difference in which the goal is not to make
difference disappear but to make it part of the political program.
Furthermore, the task requires the same breadth of global connection
as does frontier resource extraction.

—Anna Tsing (2004:211)

Talking across lines is wrought with friction, but it should be, right? Working across difference is just that: defining, embracing, and setting practices around difference, as Tsing advises, rather than erasing

FIGURE 5. Open-minded Republican at Idaho's first Climate Action Rally, Boise, October 24–25, 2015. Photo by the author.

it. People are not cut and dried in their opinions, nor are they defined by singular identities but rather by configurations of lived experiences (Bhavnani and Bywater 2009).

Working across lines of difference involves friction between people with different perspectives and friction within individuals, between their complex and seemingly contradictory views. Thus, it is a messy business, but friction, as Tsing (2004) argues, also creates productive tension. In CAIA, this was clear in how the group developed its strategy of focusing on accountability and integrity and in interviewees' stories about their perspectives. The friction of working with people who believe and do not believe in climate change led CAIA as a group to dig deeper for the roots of problems they all cared about. In individuals, changing and ambiguous opinions, particularly around opposition to oil and gas, motivations for opposing the industry, and climate change beliefs, illustrate the heterogeneity and complexity of political categories and labels (see, for example, figure 5).

Opposition to Oil and Gas

A few interviewees were immediately against drilling for oil and gas in Idaho. Most, however, were curious about the potential to be Jed Clampett, of the 1962 sitcom *The Beverly Hillbillies*—that is, an impoverished rural landowner who strikes it rich when oil begins spurting out of the ground on his property. "We thought, 'Oh drill, that's wonderful,'" Jan said. "We'll use our own resources and it will benefit the community." Economic growth was and is a message the gas industry and the state use to promote oil and gas expansion (see Idaho Petroleum Council 2016). After conducting research about fracking online, interviewees all concluded that the possibility of becoming Jed Clampett was a myth, especially for small landowners, and that the risks of oil and gas outweighed any potential benefits.

Although these conclusions drove their participation in CAIA, some CAIA members believed oil and gas *could* be extracted safely in *some* cases. "Messiness" is how I characterize these perspectives that oppose oil and gas, recognize that renewable energy makes more sense, and yet, simultaneously adhere to the belief that it might not be necessary to oppose oil and gas in *all* cases.

An example of this situation is evident in the perspectives of Luke and Brynna Smith. Both were neither for nor against oil and gas when they heard that their neighborhood would be force pooled, just two weeks after moving into their home in 2014. They wanted to know both sides of the story to formulate their own opinion, so they did research online and made phone calls to the state. They formulated an anti-industry attitude *for areas near people*. It would be fine, Brynna said, if natural gas were "out in the middle of nowhere." The Smiths recognized the need for energy and potential for energy independence, *but* they simultaneously wondered why solar could not be the focus. They also knew, through experience, what a bad neighbor oil and gas was. Luke and Brynna, therefore, were not totally against oil and gas, but mostly. They could see both sides but had, because of their research, decided to actively oppose the industry. They held views that are in friction with one another but that likely reflect the views of many people in the United States better than a completely pro or con stance on oil and gas. For example, a 2014 Pew Research Center poll found that 12 percent of Americans did not know whether or not they opposed or favored fracking (Drake 2015).

As Brynna and Luke explained:

Brynna: Well, I don't know, I kind of still feel like if they want to do this, if they want to drill and whatnot, I don't feel strongly enough against it. [. . .] I don't care that much I guess if they do it elsewhere, like out in the middle of nowhere where it's much less likely to affect people, but I think anywhere near homes, *anywhere* near homes or waterways or anything like that is, yeah, and having dealt with the gas company personally, that's what gave me the opinion. You know, they're, it's all—there's nothing transparent. There's so many lies; they just treat us like crud, so, that's where—and we say that to people all the time—we didn't have an opinion going into this. Now we do because we've lived it firsthand. So, I mean, I guess *it's not something I would say that I'm completely against ever anywhere because I just don't feel that way, but, um, nowhere where people live or would be affected, which I mean, that eliminates a lot of possibilities, [laughs] so there you go, I guess that solves the problem,* but that's how I feel.

Luke: Well [. . .] I have two different [. . .] forms of thought, I am very aligned with my wife in this thing. Initially I had no opinion, but I know that we live in a world that relies on this stuff and so, sadly, until that reliance is completely gone, it is going to be sought after, so I do understand that yes they are going to look for this stuff, and they're gonna figure out a way to get it, but I am also understanding that we are told to be responsible and stewards. *Right now it's, "let's get it and get it as fast as we can and make as much money as we can," rather than saying, "Well, can we make this last or can we make this something that we can utilize every part of it, decrease pollution, and still make the neighbors happy?"* [. . .] On the other side of it, it's like, OK, we use it, can we use it to better this [local] economy that's here? You know they keep saying, "Oh it's good for the economy." All right, so [. . .]

Brynna: Show me.

Luke: Yeah, show me. Let's start establishing setbacks and making it appealing, you know. I'm looking at a gas company side of it, let's look at it, make it appealing to the people in this area. [. . .] Let's take it here, work with the local community and do setbacks, improve; maybe we can tie it to the local power grid here to reduce everybody's power grids or use of foreign oil [. . .] and truly make it a natural gas state, where it's an efficient low-footprint community. So that's ideally what I would like to see, but right now you are either one or the other [for or against], [. . .] *so right now we understand there's a need for it, but we want it done in a responsible manner* and, you know, even though we can't stop it, we may be able to control it to improve our community and the community wherever it gets shipped out to. (my emphases)

Brynna recognized that her desire to have wells where they are not near anything that could affect people pretty much meant she was against oil and gas. Luke articulated a view focused on developing natural gas so as to enhance energy efficiency and economic benefits. Ultimately, they struggled with the tension resulting from the petro-hegemony (see LeQuesne 2019) that grips the world: we all rely on fossil fuels in daily life.

Later in our conversation, Luke also recognized that the United States should have embraced solar power in the sixties, that there has not been a lot of forethought in energy production. Luke and Brynna's conversation paralleled conversations I had with other interviewees who had complicated views about oil and gas. Alma, for instance, though she was concerned about climate change, also often said that people should have the right to develop their mineral rights if they want to. They just should not be *forced* to. Joli, who was concerned about the toxic effects the natural gas processing facility could have on her dogs, was not against oil and gas and thought that if the majority of the people in her neighborhood, who had lived there for generations, wanted to drill, they should be able to. She just thought they should pay to relocate her business.

In a national political context as polarized as in the United States, it is necessary to avoid assumptions that one aspect of someone's view on oil and gas automatically means that they have a black-and-white view on the whole industry. From a climate justice organizer's perspective, understanding the nuances of people's views is also valuable for developing effective messaging and broader movements.

NIMBY: And Not in Yours Either

We don't want to say that you can't drill anywhere, but we are going to say that it can't be here; we are going to live here.

—Peter Dill, lawyer and organic farmer, Emmett, Idaho, interview

NIMBYism is another area where views are complex. Statements like the one from Peter, who believes in being a steward of creation and is therefore strongly against oil and gas and dismayed by how the industry destroys communities (in other words, not a NIMBY), are common. Both Alma and Jim entered activism after being personally affected by industry—Alma by her CAFO (confined animal feeding operation) neighbor and Jim when Alma was put in jail. Both, however, had continued giving all their energy to the fight against oil and gas even though no land had been leased near their home. Besides driving twenty minutes to areas like Fruitland and New Plymouth, where extraction is happening or imminent, they also often drove one hour each way to Emmett or Eagle to organize people there, where there were leases, but no drilling, during my fieldwork. Their continued involvement stemmed from their motivations rooted not in NIMBY concerns, but in a sense of injustice at how CAFOs, Payette County (who jailed Alma), industry, and the state do business. At bottom, even though Jim said he was fine with people making money, and Alma said she was fine with people developing their minerals, their demands that business be conducted ethically preclude oil and gas extraction. The oil and gas industry is only profitable because of the preferential treatment it receives from government and the fact that it externalizes the cost of its waste, in the form of climate change. In Alma and Jim's preferred ethical regulatory environment, oil and gas development would not be profitable. It would not be a viable business.

Shelley and Stealth Officer (a pseudonym chosen by the interviewee) also had complex ways of relating to NIMBY issues. Both got involved when they discovered seismic testing for oil and gas happening in Emmett, about twenty minutes from where each of them lives. Stealth Officer feared for his home: "It's not a wasteland here. It's a home, it's where people live, it's where people grow, you know; it's my home. I don't want it to get wrecked."[8] Stealth Officer often mentioned that he wanted to protect his place, but he also recognized the gravity of climate change and the need to protect the whole area for future generations. Shelley, perhaps because of her experience attending the 2015 Stop the Frack Attack conference in Denver with Alma, was attuned to the national scope of the problem and the potential to help others in the country by stopping fracking in Idaho. She recognized that she first got involved to protect her valley but now knows she must stop it everywhere or no place will be safe:

> I've always felt like [. . .] why would I get involved and cause myself all that stress, you know, but [. . .] it's like you just have no choice. For one thing, we are going to be directly impacted right where we live, sooner than later, obviously, because Brad Little owns all this land, our lieutenant governor, and he's heavily involved in the industry. We know he is going to be drilling,

fracking, right here, ruin this valley we live in. But I would be fighting this hard if it was just going to be in Emmett, really, because I just, *the injustice of it,* you know, just infuriates me. There are no alternatives, like you can't even move to get away from 'em [oil and gas], when you look at the scope of what it's done across the country. It's not like you can sell your land here and move somewhere before it hits because you don't know where they are going to go next, but you know they are. (my emphasis)

So, while Shelley's initial motivation to get involved was NIMBY, the knowledge she gained from organizing helped her center on injustice as her primary motivation.

Another layer to NIMBYism in relation to Alma is that she had a particular dislike for NIMBYism in people's attitudes. She attributed this to growing up poor and being made fun of for her poverty as a child. Because of this, she saw all people as deserving the same standard of living, no matter their class (which she often talked about in terms of house size). "My number one issue," explained Alma,

is that of people who [. . .] they'll be involved, maybe, if it's in their backyard, but if it's over here [where . . .] the houses that are worth $100,000 less, maybe it's OK in [that] neighborhood, and it's clearly not. [. . .] That's what gets me. [. . .] Everybody's, we are all equal; all of us are equal. I don't care if you're white, black, purple, red, pink, gay, straight, bi, whatever, we are all people and people deserve to be protected equally because they are human beings, not because of where they live or what color they are, or what race they are, or what sexual orientation they are.

To broaden NIMBY people's eyes, Alma often said, "OK, you don't want it by your house, is it OK by your mother's house? Your sister's?"

In my fieldwork, NIMBYism was notable among people learning about oil and gas for the first time. In reference to oil and gas leasing in Eagle, however, where people are wealthier and where huge homes on grassy expanses abound, there was more of a tendency for people to say things like Dottie's comment: "It's just absolutely unbelievable, so close to all those *nice* homes" (my emphasis). That people were commonly surprised that oil and gas development would occur in wealthier neighborhoods reflects an underlying acceptance of some places as sacrifice zones (see Bell 2014; Scott 2010) and the idea that it is OK to "wasteland" some places (Voyles 2015). This perspective is not lost on people who live in these areas: "Yeah, we are expendable," Luke said, "We are low-key enough that we can be overseen, I think that's how they look at it. Go do this in Eagle, see what happens. See how many people get upset."

While there are serious problems with NIMBY perspectives, these problems often dissipate as people replace their NIMBY motivation with a motivation for protecting all communities from oil and gas. Therefore, NIMBY motivations are an important entry into organizing, if veteran organizers can, through education, complicate and broaden new organizers' views about the nature of the oil and gas industry.

Climate Change: The Elephant in the Room

Alongside opposition to oil and gas and NIMBYism, climate change is a third element about which people had ambivalent and complex views. Some interviewees expressed skepticism about climate change. While climate change skeptics are routinely interpreted as just part of the problem by liberals and climate justice activists, it is not that simple.[9] CAIA members, whether they believed in climate change or not, were doing good things for the climate by fighting natural gas. In addition, just because someone is skeptical about climate change or that it is human caused, does not mean they are not concerned with pollution or cannot recognize that our social practices of consumption, what sociologist John Urry calls "high carbon lives" (2011), are not sustainable.

Peter Dill, for example, was not convinced that climate change is human caused, but he was convinced that we are not being good stewards of the earth: "I mean at the root of it, for us it comes back to, how are we doing as stewards of the earth? What we know very clearly is that our generation is a heavy polluter and we have no business doing that." He was concerned about the system-wide problem of consumption:

> We are also heavy consumers who have forgotten a very important word, and that's conservation. And we pay for it. So climate change for us is a part of this larger issue of stewardship. It's related to pollution, which is related to poor habits and things that are hurtful for us and our neighbors. [. . .] If we can think about conserving, we wouldn't have to do nearly as many of these things to support habits that are not all that great for us. And if we would simplify our lives and not be as gadget-oriented or as interested in all the things that are very power consumptive and [dependent on] petrochemicals, if we would simplify back from some of those things, this issue could go away. So yeah, climate change is a piece of it.

Peter's beliefs center pollution as a problem. They blur the falsely drawn neat boxes of liberalism and conservatism. He thought climate change was happening, citing the science, but that it might be unrelated to humans. Peter was also a lawyer, an organic farmer, and someone who believed deeply in creation. It is important to recognize the blurred and socially constructed nature of these categories. Blurry messiness rather than neat stereotypes characterizes the political views of many people in the United States (Gelman 2011). Defining neat dividing lines is in the interest of corporations. It helps industry maintain power and divide people by obscuring their common interests.

Sherry Gordon and Jim showed similar complexities in their views of climate change. They relied on data for supporting their fight against oil and gas but were suspicious of both climate change data and climate change scientists. They thought scientists could be bought. Sherry had decided that whether climate change is real or not does not matter for her work:

> I know either there is climate change, or there isn't, and I've heard of, seen in the news, things from both sides that have really concerned me, made me think these people [vocal politicians or scientists whom Sherry thinks can all be bought by special interests] are slimy, you know, just playing a game. So, I don't know what it is, if there is climate change, I'm very concerned about it. I mean, there's clearly climate change, you know; we are experiencing weather patterns that are wacky the last two or three years in particular, and I love animals and, you know, I'm very concerned about the land. [. . .] It's like what we do about gas and oil as a community, as a nation, as a world, is clearly going to affect the planet, and whether, you know, if that's going to help the climate, then it is, but it's horrible [. . .] all the effects of gas and oil I mean are so varied and that just to me takes absolute precedence [. . .] over whether I need to decide whether there's some kind of climate change. It's interesting because people are talking about climate change and [. . .] is it human caused or not? And clearly, we have weather change, and is that human caused? Well, I don't know, how can I tell? But everything we do is human caused, all the pollution is human caused, all the horrible, you know, health effects for people and the animals, are human caused [chuckles]. We just, we have to look at what we are doing, whether it affects the climate or not.

Again, Sherry's explanation bursts all categories of thinking on climate change and emphasizes pollution, like Peter's words. She's skeptical of the science, but she recognizes climate changes in her own observations of the weather. She is concerned about the effects climate change could have, if it is real, on people and animals. She was also very in tune to the need to be careful with CAIA's messaging around climate change and how that affected who would be attracted to the group. In the excerpt, she concludes that oil and gas pollution is the most urgent problem there is. She also recognizes that everything is human caused and that humans need to closely examine our actions.

Through different language, the idea that humans need to fundamentally change how we interact with the earth is an important part of what the climate justice movement advocates. Despite their climate skepticism, some CAIA members held critical views about humans' impacts on the earth and the need to simplify, consume less, and respect the earth. Climate skepticism then, is not a failsafe predictor of other environmental and justice views. It is possible to be skeptical of the science or of the anthropogenic nature of climate change while also espousing and enacting (as was clear from Peter and Susan Dill's lifestyle) a care for the earth and those who call it home.

CONCLUSION

The friction and messiness of talking across lines make CAIA a new and hopeful model of practice for radical social change. CAIA targets the same roots of injustice as the climate justice movement. In their focus on natural gas, CAIA members are automatically linked with the broad and growing global movement against the fossil fuel industry, a movement that has gained strength because of the climate crisis. The breadth of global connection that Tsing (2004) advocates seems to be growing. The next step would be for more climate justice activists and anti-extraction activists to fully accept each other, even if they sit on different sides of the political aisle, and begin working together on a larger scale.

In focusing on common values—what everyone can agree on—CAIA works amid difference for a goal grounded in justice. They recognize that they are stronger and more effective despite and because of their membership's different views. CAIA members' choices to check particular differences at the door is a strategy that allows the group to work together on what they do hold in common. By working together, members build relationships of trust that, when sufficiently strong, can allow them to begin discussing their differences and agreeing to disagree. Eventually, agreeing to disagree might lead to changes of opinion. Building relationships, especially through working on a project that people are deeply committed to because of a core value (e.g., respect, responsibility, trustworthiness, caring, fairness, property rights) is a key first step for cultivating empathy and valuing difference.

CAIA's resistance is based on the idea that people should treat each other with respect and that in business, everyone should be a winner—businesses should serve communities. The oil and gas industry violates these core values; so does government through its connections with industry. CAIA's core ideas are informed by the life experience of its members. Their experiences lead them to create practices to uphold values that they hold dear. Jim had been a small business owner since he was seventeen and was dedicated to a number of practices in his business that support win-win situations for himself, his employees, and his clients. As he explained, "I'm all about business, I don't care about how much money a business makes; they can make trillions of dollars, God bless 'em, *as long as* all parties won. But when a business has a business plan where they take advantage of other people and they use laws to protect them and to create laws that allow them to have the ability to take advantage of other people, that's a problem." Alma grew up poor in Kentucky and served in the military out of a belief in the importance of defending the values of the United States. Ever since she was young, she stood up for the underdog. Like Alma, Sherry was always someone who would put in the work that was needed when no one else would. Others, like the Dills, were driven to protect creation. They felt imbued with a responsibility to be stewards of the earth. All of these people came together around their shared understandings of right and wrong. They think people's actions should serve the good of their communities and that all people should enjoy a good quality of life.

CAIA grasped the root of the problem in a sophisticated way that was more effective than focusing only on environment or social justice. Its name, identity, and purpose are all founded on addressing the roots of many social and environmental problems. Though no CAIA members identified themselves as anarchists, revolutionaries, or anti-capitalists—and would not, I think—in their grasp of the root of the problem, they were radical. As Karl Marx noted, "To be radical is to grasp things by the root" ([1888] 1978). Though they were not thinking in terms of system change during my fieldwork, the changes they worked to advance require new systems to function: a new system of culture where care and respect are primary, and a new system of economics and politics where big business does not control decision-makers and where decision-makers represent the interests of the people. In Jim's words: "People need to start realizing what their true enemy is and that's the collusion between unethical business and corrupt government, and even get down to the roots, get back to building society, back for people who care. The bottom line is that we wouldn't have any of these problems if everybody followed the golden rule, done, it's over. We wouldn't need government to make laws because we'd be concerned about the other person."

CAIA then, is one example of bringing people together across political difference to work for change that would benefit all justice movements.

In the next chapter, I turn to young people to examine how they, in their own ways, work across lines.

CHAPTER 5

Working across Intersectional Lines

Youth Values and Relationships

Sitting in a circle in a sunny room of a small college in Berkeley, California, in fall 2015 at the first 350.org organizing summit for California group leaders, my fellow 350 Santa Barbara member, and cofounder of our group, Max Golding, raised his hand.[1] We were generating our group norms for the weekend workshop. Max added "calling in" to our list. This was the first time I had heard the term, yet I quickly realized its salience for many of the values and practices of interviewees. Calling in is the opposite of calling someone out when they say something problematic. Instead of calling attention to a problematic behavior in a way that shames the person who engaged in the behavior, a facilitator who practices calling in will call attention to a problematic behavior in a way that encourages learning and behavior change by the person in question, who is treated as an important part of the group.

This attention to learning and relationships is central to how youth climate justice activists organize in Santa Barbara, California. While they occupy very different social, cultural, and geographic positions than the activists in Idaho, they have developed similar practices to enable working across lines of difference, particularly lines of identities and passions. In line with many of today's social movements, they prioritize movement building as a core component of their work to change the world.

This chapter takes inspiration from the words of Martin Luther King Jr. (1967), who provided the following wisdom over fifty years ago: "We as a nation must undergo a radical revolution of *values*. We must rapidly begin the shift from a 'thing-oriented society' to a 'person-oriented society.' When machines and computers, profit motives and property rights, are considered more important than people, the giant triplets of racism, extreme materialism, and militarism are incapable of being conquered" (my emphasis). King's focus on values is prescient for our current moment. In the twenty-first-century United States, values, more than evidence, play a large role in shaping public perception of issues like climate change. Values then, must be a target of social movements if we are to overcome the crises King identified, crises that the wildcard of climate change exacerbates.

In the pages that follow, I examine the values of youth climate justice activists and how they embody these values. To better understand how youth's values inform their movement-building work, I develop the concept of a "climate justice culture of creation." Climate justice culture of creation builds on John Foran's concept (2014) of a political culture of creation, or PCOC.[2] Developed in the study of revolutions, PCOC illuminates how social change emerges through the creation of revolutionary political cultures that tend to be egalitarian, horizontally organized, invested in deep democracy, and prefigurative—working in ways that reflect the world that activists want to create (see Boggs 1977–78; Breines 1982; Polletta 2002). Ideologies, idioms, networks, lived experiences, and emotions form these cultures (Foran 2014).

Climate justice culture of creation (CJCOC) extends PCOC in two ways. Foran (2014) developed PCOC through meso-level historical and comparative analysis of many movements. I develop CJCOC through analysis of the values and practices of activists themselves, centering activists' own theories. This deepens the concept's capacity to foreground activists' agency and produce "movement-relevant theory"—theory that critically engages with the dialogues and questions that concern movements themselves (Bevington and Dixon 2005). Second, I employ CJCOC to understand how values and practices, coming together as culture, constitute the process of movement building. While Juris et al. (2014) examine and highlight the importance of movement building as an *outcome*, which they define as "the creation of movement infrastructures required for sustained organizing and mobilization" (329), I examine the *process* of movement building—*how* activists strive to create these infrastructures.

In the climate justice movement, activists engage in movement building to not only enable future mobilization, but more importantly, to also create the world they want to live in and the relationships they will need in that world. Activists see movement building—creating a political culture capable of sustaining, nourishing, and growing the capacity of people to transform their experiences, ideas, and relationships into action for social change, and new ways of living—as central in addressing the crisis of climate change and social injustice.

A CLIMATE JUSTICE CULTURE OF CREATION IN SANTA BARBARA COUNTY

The majority of the twenty-nine youth climate justice activists with whom I engaged were affiliated with the University of California Santa–Barbara (UCSB), the greenest public university in the United States when I was conducting research, and the third-greenest in 2021 (Princeton Review 2015; Princeton Review 2021). UCSB bills itself as a leader in sustainability and a "center for environmental movements since the 1969 oil spill" (UCSB Sustainability 2019). Indeed, mirroring the strength of community environmental nonprofits founded in the 1970s, UCSB has one of the oldest and largest environmental studies programs in the country and a robust student environmental activist community. Since 1992, students have organized the Environmental Affairs Board (EAB), a group with an officer team of about twenty people, wielding an annual budget of approximately $30,000. The university has forty student groups related to sustainability, who come together as ECOalition.

UCSB has become increasingly diverse in recent years. In 2015, it became the first member of the Association of American Universities, composed of leading research universities, to be designated as a Hispanic-Serving Institution, meaning at least 25 percent of students identify as Hispanic. It is also an Asian American and Native American Pacific Islander–Serving Institution, meaning at least 10 percent of students hold these identities. Of my interviewees, youth were the most diverse in terms of racial and ethnic identities. While they all espoused climate justice perspectives, not all called themselves "climate justice activists." Many self-identified as environmentalists and social justice activists. All had organized against extreme energy extraction and had engaged in a number of different tactics and campaigns, including protest, lobbying, awareness raising, and education. They had protested the Keystone XL pipeline, campaigned for fossil fuel divestment, participated in California anti-fracking and climate marches, and worked on Measure P, the 2014 ballot measure to ban extreme oil extraction in Santa Barbara County. In addition to the environmental groups interviewees organized, a number also participated in an array of society- and culture-focused groups. For many, their involvement sprung from learning about environmental and social problems in high school and college classes and student groups.

Once youth activists became involved in climate justice organizing, they quickly began learning from their peers in the movement. The most important thing they learned was that values that prioritize inclusivity are vital to growing organizations and movements. As youth explained their approaches to me, four values rose to the top as ones they find effective for building and sustaining their movement. These values of relationships, accessibility, intersectionality, and community, when added to the practices through which youth enact these values, constitute a culture focused on creating a better world—a climate justice culture of creation.

Relationships

Sitting across from me at a table under a brilliantly blooming red coral tree in May 2015 at the University of California–Santa Barbara campus, Madeline Stano, attorney for Center on Race, Poverty, and the Environment, said, "If we're really trying to build climate justice [. . .] that includes how we interact and are in our personal lives." Many interviewees stressed the value of relationships, not only because relationships of trust and friendship build strong and effective movements, but also, because they are key to the *creation* component of youths' climate justice culture. Just relationships are foundational to the world they want to live in.

Fossil Free UCSB, and one of their core organizers, Theo Lequesne, learned the importance of relationships through experience. From the United Kingdom, Theo had organized his first fossil fuel divestment campaign at the University of Warwick. There, his student group had succeeded in passing, with 75 percent of the student vote, a referendum on divestment motion. Since coming to UCSB for his master's degree, Theo had struggled to retain members in UCSB's Fossil Free campaign.

To strengthen their group, UCSB Fossil Free sought guidance from Emily Williams, then California Student Sustainability Coalition (CSSC) campaign director for Fossil Free. She facilitated team-and-relationship-building workshops so group members could learn their strengths, capabilities, and knowledge levels about the campaign. This strengthened their group. Theo then recognized that at Warwick, he had organized with his group of friends and that their friendship base had facilitated their effectiveness at organizing. They had been engaged in relational organizing, which, Emily explained, is rooted "in having these really intentional one-on-one conversations that are not about campaign strategies. It's not talking about the campaign at all; [. . .] it's focusing on getting to know someone—what's, what's driving them, where they are coming from—and using a whole series of them [conversations] to build up this trust between two people. And so, when they organize together, there's that relationship built." Fossil Free UCSB's relational organizing paid off. On May 19, 2015, following the Refugio oil spill near UCSB, they held their biggest protest of the year with mock UC representatives shaking hands with oil companies as a pipeline spilled fake oil on students. Two years later, on May 11, 2017, Fossil Free UCSB celebrated a major victory when UCSB's chancellor, Henry Yang, following a four-day sit-in of his office by student activists, endorsed fossil fuel divestment. He was the first University of California chancellor to do so.

Relational organizing was a thread throughout organizing at UCSB and my other research sites. Rob Holland, who was a member of multiple organizations and had led UCSB California Public Interest Research Group's (CALPIRG) anti-fracking campaign, felt that there had been much better cohesion and mutual support among environmental and social justice/cultural groups in 2014/15 than before. He explained that the Black Student Union had supported campaigns by CALPIRG (which often had environmental campaigns) and United Students Against Sweatshops and how many student groups were recognizing their connections as they mobilized around big issues like Ferguson and #BlackLivesMatter, an important campaign for a number of interviewees. Rob thought friendships were key to this solidarity: "I think the leaders of the big environmental groups on campus are good friends with the leaders of the cultural groups, and so they just like know that it's stuff that they should support."

Interviewees recognized that, when lacking friendships, expanding social networks was a necessary first step to fostering the conditions for effective organizing. Colin Loustalot and Alex Favacho, who met in 350 Santa Barbara, lamented what Alex called "the limits of our white networks." They identified the tendency to create social networks with people whom we perceive as similar to us, the principle of homophily, which creates particularly strong racial divides (McPherson, Smith-Lovin, and Cook 2001), as a major barrier for their climate justice groups and the health of communities and social relations more generally. Colin explained: "We have this way of compartmentalizing ourselves with others that then gets reflected in our work, and it's really hard to work against that. It oftentimes feels artificial and becomes this, like, do you do it even though it feels artificial?" Colin was concerned about the possibility of tokenizing people as an outcome of trying to broaden social networks. For him, social relationships—emotional connections—were necessary for changing people's preformed ideas about the world. Culture, not facts, changes minds. In Colin's words, "You spew facts all day and you are unlikely to change anyone's mind about anything; the facts sort of help reinforce your beliefs. [. . .] People have these preformed ideas, and it's really hard to change those unless you have some kind of emotional, personal connection that will get them to, that calls on their empathy and these more [. . .] unconscious parts of them that inform their world view."

Though Colin, like others, was still "baffled" about how to expand his networks in an organic way, he was able to identify an example of when he could have done so across political lines. In February 2014, he and other members of 350 Santa Barbara held a protest against the Keystone XL Pipeline. They held "NO KXL" signs at the Santa Barbara Film Festival, hoping their signs would make it into the media's photos of stars like Oprah. A man came up to him and told him, "You are all idiots, you're ignorant," and then started "grilling" him on facts. Neither Colin nor the man budged on their positions. Perhaps reflecting his observations of a fellow organizer, who was known for her relationship-building skills and having coffee with people to build relationships, Colin remarked:

> In retrospect, after it was too late, I was like, "I should have asked him to coffee," because [. . .] I think those are opportunities where, instead of fighting and destroying to argue your point to win, those types of moments, if you can somehow turn them into something more, or you can find the common ground, I think that that's where a lot of the work can happen that's behind the scenes type work. And it happens on a social level, so when people just have empathy and give a shit about each other, then they start to listen to each other more.

Colin reflected on how turning an aggressive and confrontational interaction into a learning experience for building social relationships across disagreement and developing empathy would be an effective approach to organizing.

One way to demonstrate empathy was to "show up," a practice advocated by climate justice activist organizations (e.g., Curtis 2015). Kiyomi, who was dedicating her activism at the time of our interview to building community as president of the Board of Housing Cooperatives, stressed how reciprocal relationships between friends and showing up to each other's events can build community.

She suggested that activist groups develop more "inviting energy," and communicate a sentiment of "come participate with us on this." Through building community in this manner, she felt the diversity of the movement would grow as more people from different sectors of the community learned about opportunities to be involved. Part of the power of showing up is that it puts people in face-to-face interaction, where they can, as Rob Holland described it, "see your face and your passion," something missing in much social media interaction and online campaign actions. It also allows storytelling, an important part of social movement mobilization (Polletta 2006). As Kai Wilmsen, former leader of the Environmental Affairs Board (EAB) explained: "I think there's so much power just around sharing stories and being able to like sit down with people as people and being like, 'Hey, tell me about yourself, tell me about the work you are doing,' um, like showing solidarity when you can."

Building relationships for the sake of relationships is something interviewees identify as critical to their movement-building work. Relationships promote trust, empathy, and awareness of how and when to show up in solidarity with other movements; they also make organizing enjoyable. While it may seem self-evident that getting to know people is important for working together, not all organizations prioritize this. In chapter 7, I describe the Measure P campaign, in which a lack of relational organizing weakened the movement.

In sum, youth activists prioritized relationships to strengthen organizing work and create long-lasting friendship networks critical to building a just society within climate crisis. To cultivate relationships, especially beyond individual networks, interviewees prioritized accessibility.

Accessibility

Calling in, the concept I highlight at the opening of this chapter, is key for creating inclusive movements. It is about the ability to talk across lines, whether based on identity, political affiliation, Facebook circles, or norms of interaction. It facilitates accessibility in groups through recognition that intersectional organizing for justice is a learning process and that individuals with the best intentions can make mistakes and learn from them. It is one tool to hold people accountable when they do something oppressive, and to get them to change that problematic behavior. In a blog post titled, "Calling IN: A Less Disposable Way of Holding Each Other Accountable," writer Ngọc Loan Trần (2013) explains: "I picture 'calling in' as a practice of pulling folks back in who have strayed from us. It means extending to ourselves the reality that we will and do fuck up, we stray and there will always be a chance for us to return. Calling in as a practice of loving each other enough to allow each other to make mistakes; a practice of loving ourselves enough to know that what we're trying to do here is a radical unlearning of everything we have been configured to believe is normal." Calling in facilitates relationship and friendship building across lines among people working for justice. Interviewees' efforts to call each other in were key to making organizing accessible, to "meeting people where they're at."

Kai explained "meeting people where they're at" as follows:

> I think the most important thing [for facilitating conversations about environmental justice] is to not like hit people over the head with it. [. . .] You have to be patient. [. . .] I mean like with any conversation, because I believe it so deeply, I just try to *meet people where they're at* first and [. . .] then have a conversation as individuals and be like, "Hey, I hear what you are saying, but what about this?" And just like, not being like, "You are wrong and this is the right way to do it," but kind of like, offering different perspectives and seeing what they think of those. (my emphasis)

Kai's explanation highlights a number of practices and perspectives that interviewees identified as key for effective organizing. As Kai communicated, the concept of "meeting people where they're at" requires patience, respect, and listening. Echoing Kai's sentiments, Kiyomi advised:

> Try to befriend as many perceived enemies as possible [laughs], and just actually become friends and hang out and just be like, "I feel this, this, and these are the facts, how do you feel about this?" You know? [. . .] So in a way you have to be, not passive, but gentle, and be patient with the conversation, instead of being like, "I'm this blah blah blah blah, this is what I believe, BOOM," they're going to be like, "Who are you? What?" and not even process anything that you have to say. Connection, man, it's powerful [laughs].

By recommending activists ask people how they feel about the facts and be patient and gentle, Kiyomi alludes to the importance of feelings and values for informing how people interpret information. Her hunch is widespread among interviewees and consistent with literature on climate change beliefs (McCright et al. 2016). In the eighty-seven studies that McCright et al. (2016) review, scientific literacy is the second-*least* important predictor of pro-climate views. Most of the time, the facts are not enough to convince people. Through experience, interviewees had learned that appealing to values, strategic framing, and friendship, or preexisting networks (McAdam 1988), were also vital.

Kyle Fischler, treasurer of CSSC and former EAB leader, stressed how educational events that were celebratory and inviting to all kinds of people could be employed to facilitate understanding across different worldviews: "If [someone is] not already partially on your side, you're not going to get them on your side by writing a very strong case against their worldview or what they believe in. [. . .] In order to get all those people that are on the fence, I think some of the most effective ways of organizing are just hosting celebrations, where you can also ingrain some education into it." A good example of this, according to Kyle, is the annual Earth Day celebration in April in Isla Vista (the town where many UCSB students live). Hosted by the Environmental Affairs Board and funded by student fees, the event showcases local bands, food, and gives student groups opportunities to share their information with attendees. Another way to accomplish nonconfrontational communication with people on the far side of what activists call "the spectrum of allies"—that is, the spectrum of support or opposition, where people on the far side of the spectrum are actively oppositional to a cause and those on the near side are actively supportive—was strategic messaging (see Moore and Russell 2011:49).

Theo, who completed a master's degree focused on climate communication the year after our interview, was studying how to talk about climate change in ways that appeal to audiences beyond leftists and progressives. He thought building broad support for climate justice was critical and depended on appealing to values: "I think that threat has to be communicated more successfully to appeal to values of people who—you know, I don't like the phrase mainstream, but sort of—who are influenced by dominant culture. And I think if that vast group of people can be brought to understand, or at least [. . .] to kind of see climate justice principles as legitimate and the sort of neoliberal solutions as illegitimate, that's when the climate movement starts winning. I think the only way to win is with a climate movement that is inclusive." As Theo highlighted, communicating the threat of climate change by effectively framing climate change, is critical to inclusivity and accessibility, both key to winning. He argued that this, however, could be accomplished without saying "climate change." Instead, one could talk about the drought in the Midwest, or extractive industries poisoning water and food. Communicating these messages and creating interactions that enabled effective communication, interviewees argued, was best accomplished through one-on-one, face-to-face conversations—through relationships, as described previously.

After communicating the threat, Theo stressed the importance of communicating alternatives to the present system, of imagining a different future. Kelley (2002) and Pellow (2014) eloquently argue for the importance of visioning. For many interviewees, a better world is an inclusive world that values difference and recognizes interconnections. To begin realizing this type of future, interviewees prioritized creating intersectional movements grounded in individual recognition of how the intersections of identities inform experiences of privilege and oppression.

Intersectionality

Youth climate justice activists are acknowledging and embracing difference, exploring intersectionality, in sophisticated ways. First introduced by Crenshaw (1989), intersectionality is the recognition that multiple identities intersect to inform an individual's lived experience of oppression and privilege. Activists recognize intersectionality on the individual and movement level, working to recognize how all of their issues and movements intersect and are rooted in the same structures of inequality (see figure 6).

Take, for example, the California Student Sustainability Coalition's (2014) statement in solidarity with Ferguson. Interviewees Emily Williams and Unique Vance were among the coauthors:

> We recognize and affirm that our struggle and liberation is indelibly bound to the liberation of others. We cannot have climate justice or a sustainable planet without racial justice. [. . .] Our movements must be intersectional because our lives are, our very identities are. [. . .] The injustices that Ferguson faces today are rooted in the same injustices that we fight here in California. Though we all feel their impacts differently, we have a moral duty to highlight these impacts and those who are the most vulnerable if we are to find justice. Most importantly, we need to recognize that we are part of the same struggle. If we want freedom from the fossil fuel industry, if we want freedom from tuition hikes, then we must also have freedom from oppression and racial injustice.
>
> CSSC must stand with Ferguson in order to—together—resist these injustices and to—together—build the future we want and need to see. We also recognize that we are part of a larger community that holds a LOT of privilege, and although our membership and leadership is by no means monolithic, the very point of entry (a university or college) is from a place of privilege. Our struggles may or may not be the same but we are bound nonetheless. (California Student Sustainability Coalition 2014)

FIGURE 6. UCSB climate justice activists march in the May Day Parade in Santa Barbara, California, May 1, 2014. Some hold a "System Change Not Climate Change" banner; others hold LED signs that say "Stop SB Gang Injunction," an injunction targeting Latinx youth. Immigration reform was another focus of the march. Photo by the author.

While interviewees' grasp of intersectionality varied, all recognized the importance of connecting social issues to environmental and climate issues and the shared roots of all types of injustice. Simultaneously recognizing how struggles are "bound" yet distinct makes possible broad coalitions that bring together people who care about different issues, not to erase difference, but to show how different struggles can all be improved by working together toward common goals. CSSC's statement not only recognizes the need for solidarity among movements fighting all types of injustice, but also, importantly, the privileged position of college student organizers.

Activists' awareness of their own privilege emerged in the course of interviews. For Nia Mitchell and Unique Vance, both self-identified women of color and chair and former chair, respectively, of EAB's Environmental Justice working group (among many other activist roles), their privilege as Americans was important. Privilege informed Unique's analysis of the crisis and motivated Nia's activism. Nia's family came to the United States from Cuba. With the privilege she enjoyed having been born with US citizenship, as compared to some of her family who still resided in Cuba, Nia felt a strong sense of duty to "give [herself] to the movement." She explained:

> I have to feel like what I'm doing is helping someone, whether that's the advancement of the community that I come from, or just like, the global majority in general. [. . .] I think that's why I do it. [. . .] I learn way too much sad shit on a day-to-day basis not to act, and not to be proactive, and not to constantly dismantle. [. . .] It's just in my blood, I really feel like it's [. . .] the whole reason I'm here, almost, is to like help change something you know and commit my life to struggle in that way, because I'm lucky enough to, because I don't have to live how [the people affected by] what I'm learning about live, you know what I mean?

Interviewees who identified as white men were also self-aware of their privilege and recognized how it could be used strategically for organizing. Theo felt the divestment campaign was where he could use his privilege as a college student most successfully. As a student, he had a stake in and ability to influence where university endowments, a primary target of the divestment movement, are invested. Arlo, a member of 350 Santa Barbara who helped spearhead the Measure P campaign at UCSB, struggled with feelings of guilt about his privilege as an educated white man and over how this could inform his actions. He saw potential to leverage his privilege to spread the movement's message. In his words: "There is so much privilege and like history that I am benefitting from, and it's really hard not to let the guilt from that like overwhelm me and cause this sense of shame. [. . .] I feel that I really need to be talking to some folks who might not listen to others or who might [. . .] really be tuning out this climate change thing, but you know, seeing this unfortunately young white male who just graduated college, they might be more receptive to hearing that."

Another way awareness of privilege manifested among white men was in their commitment to feminist values of gender equality and recognition of intersectionality. Two of these interviewees mentioned that they had taken on the responsibility to explain the significance of feminism to their non-activist male friends. When I asked Kai about whether he had friends involved in activism, he said many of his friends were not activists. He recounted the following conversation:

> We were discussing if feminism is necessary, if we need feminism, and I was like, "Oh my God yes, like why, what!?!" And we had this whole long

conversation and like we didn't really get anywhere; we were just going back and forth. But then a couple days later he [Kai's friend] texted me and he was like, "You know, I've been thinking about what we talked about, and I think I understand now what you were saying about like the privilege of being a man and how women are oppressed." And I was like whoa! [. . .] I was a little bit shocked too, because he is more on the conservative end.

Kai, an ecology major, had learned about feminism and intersectionality from student leaders of his environmental group. His social media feed, coming from his many "socially conscious Facebook friends," and online articles, also shaped his understanding of these concepts.

Exploring privilege and inequality allowed interviewees to begin to craft themselves and their organizations in ways that challenged stereotypes and empowered participation from all group members. Michael Fanelli reflected: "The people that are educated enough and invested enough to become involved in this sort of thing [activism] are people that, you know, are aware of these [inequalities and stereotypes]. [. . .] I don't know how to describe it. [. . .] People aren't close minded, people understand that [. . . stereotypes] aren't like legitimate things and that they're just based on, you know, old traditions." As this quote hints, activists identified problems with traditions, practices, and perspectives of dominant culture. Making change begins with identifying problems and then moves to creating alternatives. Thus, understanding privilege and inequality had to do with activists' work to identify the barriers to the world they wanted to see. While intersectionality is almost a buzz word in some movement and academic circles, it is relatively absent from social movement theories (e.g., Snow, Soule, and Kriesi 2007).[3] It is crucial for developing an understanding of the climate justice movement because understanding, prioritizing, and putting resources toward articulating how the movement is intersectional is the first step to achieving a movement of "everyone." As scholars of collective identity and framing have noted, people have to see themselves as part of the movement or agree with movement messages to begin engaging or to remain engaged. The climate justice movement must, therefore, articulate how climate change is also about racial, gender, and economic justice if it hopes to gain support from these other movements.

Just as activists worked to prioritize intersectional understandings of themselves and movements, by recognizing privilege, so too they strove to prioritize community by recognizing the factors that discouraged it.

Community

Calling in to conversations and relationships across difference meant overcoming a culture of not talking to people with different perspectives, a culture of division and "anti-community." Rob was very concerned with lack of communication in US culture. When I asked him what he hoped for the future, his response paralleled the organizing practice of agreeing to disagree employed by activists in southwest Idaho:

> I hope for a reversal of I feel like the current trend of anti-community and like anti-talking to one another. [. . .] I feel like we used to be able to, I mean like before we had ways to communicate like so easily with people that were like-minded with us, [. . .] we had our community and like our neighbors and our mailman and stuff. They were there, and you were forced to talk to them 'cause they were the only people around, and if you disagreed with them [. . .] you respected them; [. . .] you knew that they were there to talk to and hopefully you could still count on them even if you disagreed with them um, but [. . .] we now have [ways] to like not talk to people really, or to have our niche groups to where we don't have to talk to people that don't agree with us.

Rob went on to apply this observation to his own experience of organizing, where he noticed a lack of student receptivity to discussing campaigns with strangers: "We think that people like coming up to us in the Arbor [popular tabling and lunch location at UCSB] and asking them if like they know anything about a campaign is weird. [. . .] And we think somebody calling us to talk about a campaign is weird, and I think that's really detrimental to any organizing, if we can't talk to strangers and like feel that people are in a community trying to make things better and listen to one another." Other interviewees had also experienced "anti-talking to one another" in their organizing and felt that social media and technology exacerbated the problem. As Sarah explained, "Especially nowadays, everything being texting and Facebook on the computer, like it's awkward for people to call each other, which is sad."

For two activists with 350 Santa Barbara (an off-campus group), this dearth of community communication manifested in divisions between Santa Barbara's environmental and Latinx communities. Arlo thought Measure P's failure to garner Latinx support was "one symptom of this system that really divides us up as much as possible and makes sure we're not talking to each other, [. . .] that has set things up so that we are not supposed to be talking to our neighbors who may not look like us, that are dealing with the same things that we are."[4] When phone banking for the 2014 Measure P campaign, Alex Favacho talked to people on both sides of the political divide in Santa Barbara County: "Some phone calls could be actually really nice, someone's like, 'Oh yeah, yeah, of course I'm going to vote for that,' and others are like, 'I own guns and stuff like that, stay off my property.'" This divide was informed by the longtime presence of the oil industry in northern Santa Barbara County, a characteristic that created a sharp contrast in livelihoods and politics, with more oil jobs and conservative political representatives in the north and more tourist sector jobs and liberal politics in the beach towns of southern Santa Barbara County (see chapter 7 for more on this divide). Political sorting like this, where people tend to live in communities with similar political beliefs to their own, is increasingly common in the United States (Bishop 2008). Such stark divisions and resulting lack of experience communicating with people holding different views likely contributes to the "toxic" (Hoggan 2016) state of discourse in American politics. Alex and Arlo, as well as other organizers, came away from the Measure P experience convinced that climate groups on and off campus had to be much more committed to justice, specifically issues such as health and economic justice, to gain support from the Latinx community in Santa Barbara County.

Talking across divides is no easy feat, however. It can be quite frustrating, interviewee Maria Castro explained, because again, facts alone often do not change deeply held beliefs that can be rooted in participation in certain occupations, like the oil industry. As Kori Lay, a former leader in EAB said, "people don't want to think they are wrong." Besides challenging viewpoints, just getting people to listen was also a challenge. Though Michael felt that facts do speak for themselves, because "[when you] are able to rationally explain these things, there's not much convincing that needs to be done," he qualified this belief with: "When someone is actually willing to listen to you." Interviewees, however, thought young people on college campuses were more receptive to learning than the general population. In Kori's experience, "campus is a really great place [. . .] because I think people are more likely to change their opinions on campus, because they are students there, and they're there to learn."

To address these problems, interviewees thought that the culture of communication and community involvement needed to change. Non-communication, lack of listening, and refusal to consider new information were all commonplace in the lived experience of their activism.

There was a general desire for a lifestyle where people talk to people they disagree with. For Rob, this would be a lifestyle of: "not like defriending somebody on Facebook immediately because they have a different view than you, of not avoiding somebody you see on the sidewalk because you know they are going to ask you about how you are voting [. . .] to be more engaging with people. [. . .] I guess what I hope for the future is a more community-oriented society, that people will feel more empowered to work for what they believe in, which I think will create more environmental good."

Like Rob, Emily advocated for changing culture, but she highlighted activist culture. She hoped activists could be more inclusive, especially of frontline communities—communities most affected by climate change and fossil fuel extraction—and could focus on campaigns that targeted cultural change. Changing culture, according to Emily, was, "a way more powerful intervention point in the system, rather than just changing a law." Making these cultural changes toward community within and beyond activist groups was a priority for all interviewees. While relational organizing is a way to build community, community as a value is broader. It integrates a belief that people should be willing to have explicitly political conversations and be open to beginning relationships with neighbors.

Putting Values into Practice

Activists' values translate directly into and are informed by movement-building practices. Relational organizing and calling in are both values and concrete ways to build movements that center relationships, accessibility, intersectionality, and community. They occur in the context of other practices and participatory democratic forms of organizational structure that are horizontal.

The majority of the groups that interviewees organized with have broad leadership structures. The Environmental Affairs Board (EAB), for example, had twenty-one officers with two cochairs in the 2015–2016 school year. Students Against Fracking began with three cochairs but rotated facilitators during meetings, using consensus decision-making to organize campaigns. During this research, 350 Santa Barbara had no leadership structure or positions. These horizontal leadership forms are one way groups tried to empower leadership from diverse members.

While groups are open to all, success in attracting diverse membership was mixed. Interviewees were aware that many of the small environmental groups tended to be made up primarily of white, hetero-sexual, gender-normative students. To them, this illustrated a lack of diversity. They were conscious that they had much work to do to realize the values at the heart of their organizing. However, they also high-lighted the diversity of perspectives, majors, and backgrounds within these groups' membership and noted that certain groups were striving for progress on diversity and inclusivity through values outlined in this chapter. In particular, interviewees gave EAB good marks on diversity. Over half of the officers of EAB during 2014–2015 and 2015–2016 were students of color. This matched the proportion of students of color at a forty-person EAB general meeting I attended in spring 2016.

At the interactional level, activists developed organizing structures to support inclusivity and learning about inequality and privilege. Interviewee Elie Katzenson, who had organized with multiple groups, recounted how, to be inclusive of gender diversity, organizations like EAB and 350 Santa Barbara sometimes asked each participant to tell the group their preferred gender pronoun, typically they/them, she/her, or he/him, during introductions. Emily Williams really liked "the culture of just naming things as they come up, so saying when something is sex-ist, saying when something is racist, just calling it right out especially in organizing space, having more workshops on anti-oppression or actu-ally discussing what diversity means." A different approach to this prac-tice, as introduced earlier in the analysis, is "calling in," which is also expressed with the phrase "ouch, oops." "Ouch, oops" was, like calling in, a norm Max Golding first introduced me to where, if someone says something offensive, people feeling the offense should say "ouch" to signal the hurt it caused them. The person who said the offensive thing would then recognize their mistake, "oops," and change their behavior in the future. It was a less confrontational—more relational—way to teach each other about inclusive language than "calling out."

In Unique Vance's view, calling out was not only more abrasive, but also less effective than explaining why people's positions are problem-atic. Along these lines, Unique cautioned against "playing the Oppres-sion Olympics." She thought it was important to take into account all of the different sectors of people's lives to recognize different forms of privilege and oppression, and most importantly, how they all connect with capitalism, how different groups of people share many of the same experiences. Insights like these helped student organizers develop their own diversity trainings and organize statewide convergences to spread good organizing practices within the movement. These practices were bolstered by their commitment to relational organizing, which, in Emily Williams's words,

> goes back to [. . .] actually spending a lot of time talking to people who might be perpetuating some of the oppression that we see in organizing circles and finding out why they are doing it, what experiences have they come through that's told them that that's OK, and recognizing that that's not them as a person, that's how society has molded them. And working with them that way to show that you still trust them, you still respect them, you still care for them, but, yeah, it's, we all need to work on getting better.

Interviewees also felt groups could improve their outreach to diverse students by clarifying the connections between social and environmental issues. For some, like Unique, capitalism was a clear way to do this: "Climate justice is when you tackle the root causes of climate change, which is capitalism, you know, bases of exploitation [. . .] so you attack the root of the issue." However, not all interviewees or groups were attuned to anti-capitalist critique. According to interviewees, Fossil Free and CSSC were doing good work to connect environmental, climate, and social justice. They were not challenging capitalism but were challenging its most damaging qualities, for example, profit without conscience and externalities, in the case of the divestment movement.

In contrast, Kori Lay felt that the EAB, despite the diversity of its participants, still had a long way to go to clarify these social and environmental justice connections, explaining:

> Outside of that [diversity training for EAB officers] [. . .] I don't think it's [social justice] addressed as much. [. . .] Multicultural groups on campus still kind of look at us [EAB] and think of us as like a white person club because of the environmental movement, the way it is still, and also just because I think we are still kind of portrayed as like, we're just trying to save the trees, like we don't have the right face on yet, so I think we're still not as inviting as we should be. I think that's partially because we don't talk about social issues enough in the club. We do have an environmental justice chair, which I think is a good position to have because it does show that we do care about those things and that's something that we have like campaigns on and things like that, but I think it's still not loud enough in the club.

Kori had worked to facilitate CSSC's transition to focus more on social justice when she was lead organizer of the yearly CSSC convergence. She told convergence panel organizers that at the most, only a third of the panels could be purely environmental focused. She invited a keynote speaker who told attendees: "Hey, you know what, right now we are on Native American lands that we took away from them." Kori recounted, "like straight out of the door, that's what she said, and [you] saw the room was just kind of in shock, kind of taken aback, which is what I wanted, I wanted people to hear and then talk and be uncomfortable about it. [. . .] I really wanted people to have hard conversations."

Fossil Free had taken up social justice explicitly by linking their campaign for fossil fuel divestment with frontline communities. Theo explained that figuring out how to communicate the social justice aspects of their campaign, and how Fossil Free is a solidarity campaign, could strengthen their message. He went on to lament the general separation of environmental and social issues, arguing for the need to clarify inter-sectionalities: "On UCSB's campus [. . .] it is so ingrained [. . .] the idea that the environment is separate from other justice organizations, and that's something that different organizations need to work on and sort of see the intersectionalities of them."

Horizontal structures, efforts to be inclusive and attract diverse membership through meeting practices and campaigns focused on social-environmental intersections, and relationship-based organizing characterized interviewees' activist groups. Activists' values both facilitate and are embodied in these practices. Both undergird the climate justice culture of creation at the heart of movement building.

THE SIGNIFICANCE OF A CLIMATE JUSTICE CULTURE OF CREATION

Climate justice movement building occurs through developing and practicing justice-oriented values, values that shape and are shaped by the broader political culture of activists' organizing. Foran (2014) posits that idioms, ideology, networks, lived experiences, and emotions can sometimes come together to form political cultures of creation capable of mobilizing revolution. The approach is useful because it argues that social movement culture is a fluid configuration of components that play varied roles depending on context. Working to identify the character of each can give a fuller picture of a social movement culture. Youth activists were drawing on all of these components: idioms or expressions, like climate justice; ideologies such as feminism; and their lived experience as college students coming of age during climate crisis. They were constructing networks grounded in friendship and recognizing the importance of emotions of passion and joy for building movements. These elements of Santa Barbara youth's climate justice culture of creation (CJCOC) are expressed in particular and localized ways in the values they share.

Understanding activists' values not only illuminates threads connecting different elements of a movement culture, but also deepens understanding of the variables that are the focus of social movement scholarship, including collective identity, strategies, tactics, and frames. The values of community and relationships, for example, motivated youth on and off campus to approach strangers to talk about politics. These inspired them to communicate the passion they felt for their work and to "show up" at each other's events. Values of intersectionality and accessibility informed activists' methods for structuring their organizations horizontally and hosting events and interactions on different topics and at different levels—sometimes mostly for fun. In other words, all four of these values informed how interviewees built collective identity. Understanding this process is not enough, however, to gain a full picture of the youth movement in Santa Barbara. Strategies and tactics are also important and also informed by values. Valuing intersectionality and privilege led students to recognize their leverage in the divestment movement and informed why they prioritized not only divesting, but also reinvesting in the communities most affected by energy extraction. Finally, all of these values inform the messages or frames that activists employ to mobilize support for their work.

Youths' values and actions are not new or invented by these activists but have been built overtime by many movements (Polletta 2002). Horizontal, or participatory democratic forms of organizing have been practiced by student movements, labor movements, women's movements, the global justice movement, and community and direct-action campaigns (Doerr 2007; Polletta 2002).[5] They are strategic because they help activists develop skills, solidarity, and innovation (Polletta 2002). Relationships have been an important building block of participatory democracy in these movements; the feminist movement of the 1960s and '70s was particularly committed to friendship and egalitarianism as foundations of organizing (Polletta 2002). Similarly, all four values that I analyze resonate with Juris et al.'s account of activists at the US Social Forum, who highlighted the importance of fun and energizing events; building "deep" and "genuine" relationships; and constructing multi-issue and multigeneration coalitions for movement building (2014:335–8). The very emergence of the World Social Forum, a "movement of movements" that prioritizes values and goals of diversity, inclusion, participation, and intentional relationship building, depended on similarities in "norms, practices, and values" among movements in different parts of the world (Smith and Doerr 2016:342–43). My research echoes findings of these scholars, suggesting the translation of values among different movement contexts. In my research, activists encounter these values, part of what Tsing (2004) calls "activist packages," in organizing, academic, and social media spaces. They then adapt and apply them to their specific organizing contexts.

While these values are not new, it is significant that young people in the climate movement are practicing them. The larger environmental movement that the climate movement is part of has not historically prioritized social justice (Taylor 2016). My research demonstrates that members of the climate movement are taking social justice organizing principles (see Southwest Network for Environmental and Economic Justice 1996) to heart, providing individual-level evidence that change toward justice is occurring in segments of the climate movement. This type of change is critical to building a broad-based, inclusive, and powerful movement. In addition, part of what informs people's choices about which organizing methods to use is not only how effective they think something is, or how it resonates with their ideology, identity, or things they like, but what is familiar (Polletta 2002). The fact that youth climate activists are getting their first taste of organizing within a justice-oriented participatory democratic style, that they are simultaneously crafting based on little previous experience in hierarchical or status quo groups, bodes well for achieving climate justice goals.

Alongside positive outcomes of this CJCOC, there is also potential for exclusion. Like broader environmentalist culture, this CJCOC risks appearing as if it is only for people with certain interests, backgrounds, identities, and bodies (see Krueger 2019). The culture itself could set social boundaries and replicate racial, gender, age, and class biases. For example, it is possible that a working-class person who shops at Walmart, eats beef, and listens to country music may feel a lack of connection to this CJCOC. However, I think the willingness of youth activists to meet people where they are at and build relationships has potential for overcoming such barriers. Many students at UCSB who are supportive of climate justice are working-class and come from cultural backgrounds with diverse environmental norms (e.g., around meat eating, organic food, or composting). Student organizers could build on and strengthen relationships with students from these backgrounds to understand how to overcome the risks of exclusive cultures.

Finally, understanding values can also help scholars and activists understand what activists define as "successes." For youth, successes came in the form of Kai's friend understanding the importance of feminism, or the people on Rob's dormitory hall supporting Rob on the night of the election when the ballot measure on which he had worked so hard failed. As Rob remembered, with a smile on his face: "Even the ones that disagreed with me were very [nice and sorry for me], [laughs][. . .] they knew how much I hadn't been on the floor [of the dorm] because I had been working on Measure P." Rob, in other words, had built community among people who did not agree with his politics, creating connections that could pave the way for future expansion of the movement.

Fossil Free UCSB went from struggling to retain members to having ten core organizers. It also, along with the statewide Fossil Free coalition, experienced a major success when the University of California divested its entire $13.4 billion endowment and $80 billion pension fund from all fossil fuel companies in 2019—the largest divestment of any public university ever (Williams and LeQuesne 2019). Through the involvement and leadership of interviewees, community group 350 Santa Barbara changed its mission from one centered on climate action to one focused explicitly on climate justice. Their "about us" page reads, "We are dedicated to building an inclusive, diverse, and empowered climate justice movement in Santa Barbara County" (350 Santa Barbara). This change occurred after many conversations in which members educated fellow members about climate justice and what climate justice may look like in the context of Santa Barbara. While this shift does not mean all actions of the group are centered on justice, it is a step in the right direction. These changes were successes because they are instances where activists saw positive effects of centering certain values in their movement-building work.

CONCLUSION

Knowing the colour of the sky is far more important
than counting clouds.
—Robin Kelley (2002:11)

This chapter argues for understanding movement building as a process of expressing and practicing values that compose a "political culture of creation" (PCOC) (Foran 2014), a culture focused on vision, rather than only obstacles—the sky, rather than the clouds. PCOCs are the "webs of meaning" (Geertz 1973) that members of a social movement shape and live within. They are constituted by the interactions of ideologies, idioms, networks, emotions, and lived experiences (Foran 2014). Activists' own theories are also important—each element of a political culture of creation is expressed on the ground through values and practices. I propose a "climate justice culture of creation" to characterize the PCOC that climate justice activists are cultivating in order to build a broad-based movement.

The youth activists in this Santa Barbara movement are cooperating to form activist groups that have horizontal leadership structures and are intentionally anti-oppressive. They recognize the interconnections of social and environmental problems and how, no matter where they come from, who they identify as, and what issues they care most about, they must engage in shared struggles for their futures. They have discovered the power of relationships, accessibility, intersectionality, and community for learning, organizing, and creating effective campaigns. Taking this knowledge, they are calling each other to the table to embrace their difference, simultaneously check and lever their privilege, and learn from each other through listening, care, and patience. The resulting relationships allow them to mobilize resources, passions, and energy to imagine and model the justice they want to see in the world. Step by step, they take action to change the political and cultural status quo for climate justice. From the perspective of movements and scholars interested in achieving social justice and a healthy environment, this is a positive sign of change occurring in sectors of a movement that has historically failed to appropriately engage with social justice.

I now shift the analysis from looking at how activists work across lines based on individual beliefs, identities, and interests to how activists work across lines based on organizational form.

CHAPTER 6

Working across Organizational Lines

*Grassroots and Grasstops Tensions
and Possibilities*

The kind of change you're talking about—anything feasible
within the current political system—really won't do us
any good.
—Tim DeChristopher, grassroots activist (quoted in Stephenson
2015:199)

If the utility needs to think that they are going to build a gas
plant in ten years in order to [shutter their coal plant], I'm
not gonna like jump up and down and light my hair on fire,
because I know I'm going to have like ten more bites at the
apple to shut down that gas plant—but what I need today
is the commitment to close coal.
—Ben Otto, Idaho Conservation League, Boise, Idaho, Interview

Movements for social change have long debated the utility of working
within existing systems. They have struggled over whether to work for
reform or revolution. Activists discuss this question within the groups
that I worked with. One central characteristic that illuminates the divide
over this question is that of organizational form—the type of organization an activist belongs to. In this chapter, I examine tensions between
grassroots and grasstops organizations.

Grasstops organizations, as I use the term, are nonprofit organizations that are structured in hierarchical ways in which a few people
make decisions and direct volunteers to follow mission statements. They

typically focus much energy on fundraising to support staff and, in some cases, have dues-paying members. In my research, activists who work for grasstops organizations are much more pragmatic, in the sense of doing what is practical within existing systems, than grassroots activists.[1] As employees of long-standing organizations with donors, mission statements, and legacies, they tend to be more invested in status quo organizing. They "accept the climate science, but fail to calibrate their response to the challenge this poses" (Rosewarne, Goodman, and Pearse 2014:5). This type of organizing stresses how much there is to lose, emphasizes incremental legal and regulatory change, and depends on having a seat at the table with decision makers and industry. It is wedded to the current system, capitalism, and does not typically look beyond this system to imagine a post-capitalist world.

The grassroots volunteer activists within these same communities tend not to feel adequately supported by these larger, better-resourced organizations—a common experience of environmental justice activists (Cable, Mix, and Hastings 2005). In fact, they often feel excluded and isolated. More of them are working to imagine and enact what Rosewarne, Goodman, and Pearse (2014) call a "realistic politics of climate change, one that meets the challenge of climate science in ways that cannot be dismissed" (6). These authors argue that a realistic politics of climate change is one that works outside conventional politics, questions the status quo, and prioritizes system change. Though activists who work in this way are perceived as radical, Rosewarne, Goodman, and Pearse point out that they, in fact, "pragmatically apprehend the challenge and seek to produce responses that have a realistic chance of delivering climate stability" (5). This surprising divide, even among people who all consider themselves environmentalists or activists, resonates with Smith's thesis: "Just as we must not presume that we cannot work with unlikely allies, we must not presume that we should always work with people who are perceived to be our likely allies" (2008:200).

This chapter analyzes the relationship between grassroots and grasstops organizations that I see as acting within, to varying degrees, the climate justice movement in Santa Barbara and Idaho. The climate justice movement works to achieve the twin goals of social justice and a livable world, recognizing that one cannot be achieved without the other. It is more explicitly justice oriented than the environmental movement and more globally oriented—because of the global scale of climate change—than the environmental justice movement. What people view as possible, the level of trust required to work together, and how people perceive different messages and strategies are some of the primary tensions between these grassroots and grasstops sectors of the movement. Members of grasstops organizations tend to take actions *because* they are feasible within the current political system. Ben Otto, in the opening excerpt, for example, does what he thinks is politically feasible—accepting plans for a natural gas power plant in order to close a coal plant. This is much different from climate justice campaigns to keep *all* fossil fuels in the ground. Actions like Ben's are actions that DeChristopher and other grassroots activists see as "not doing us any good." While there are examples of grassroots and grasstops working together, from the perspectives of activists in both sectors, these collaborations can prove more draining than beneficial. If the movement is to talk and work across lines, the dividing lines between grassroots and grasstops must be more permeable. A broad-based and powerful movement requires nothing less.

My writing in this chapter, as in the rest of the book, is informed by my standpoint as a grassroots activist. Much of the motivation for this research, and the social capital that made it possible, is based in my relationships with grassroots activists and organizations. Combined with my biography, these relationships make me an optimist about people's capacity to work together and someone who believes that achieving climate justice requires a radically different form of social organization. I do not see my vision of "success" within capitalism. In line with Rosewarne, Goodman, and Pearse (2014), I do not think capitalism is "pragmatic" in light of climate science. I also, however, recognize that legislative and political wins today, within established systems, can decrease pollution, protect communities, and award reparations to those who are suffering. I know from experience that personal environmental and social justice actions can serve as springboards for individuals to join and make efforts for systematic change. These daily actions can be small wins and progress to nourish the soul along a difficult and bleak road of system change. In sum, I acknowledge that individuals have overlapping and contradictory layers of change-making beliefs and think most people who join or work for a grassroots or grasstops organization have good intentions. Activists' dreams, choices, and reasoning are more complex than the binary characterization grassroots/grasstops suggests. I make this distinction to highlight points of divergence. Understanding these points as roots of tension, will, I hope, enable activists on both sides to transform these tensions from break points between their organizations, to bridges.

CONTEXT

The communities where I conducted research varied widely in terms of prevalence and strength of nonprofit, or grasstops, organizations. Before diving into the analysis, I therefore provide a snapshot of the nonprofit organizations with which interviewees worked and consider how this study relates to recent work on grassroots/grasstops divides.

According to the National Center for Charitable Statistics, Santa Barbara County had 809 nonprofits and 68 environmental nonprofits in 2013. There were 20.12 organizations per 10,000 persons and 1.69 environmental organizations per 10,000 persons. The *city* of Santa Barbara had 42 of those environmental nonprofits in 2013, 4.56 per 10,000 persons. This demonstrates just how concentrated environmentalism is in southern Santa Barbara County. When looking at all nonprofits, organizations dedicated to any issue, the number for the city of Santa Barbara jumps to 473 organizations, 51.30 per 10,000 persons, with per capita expenses of $23,781—highlighting the wealth dedicated to philanthropy in the community. In 2000, 39.60 percent of households with people over twenty-five years old had college degrees.

Idaho field sites varied tremendously in nonprofit concentration. The city of Boise, which houses the offices of the largest environmental groups in the state, had 36 environmental organizations, 1.94 organizations per 10,000 persons, in 2013. They had $106 per capita expenses. Payette County, where all natural gas production was occurring as of 2017, just one hour west of Boise, had zero environmental organizations. In the City of Fruitland, where forced pooling within Payette County is concentrated, only 8.23 percent of households with people over twenty-five have a college degree (in 2000). In contrast, the city of Sandpoint, Idaho, where oil trains bottleneck on their way to the West Coast, had 4 environmental organizations, which comes out to 5.79 per 10,000 persons. Their expenses were $80 per capita. When considering all nonprofit organizations, the city of Sandpoint has 46! This means there are 66.54 organizations per 10,000 persons with per capita expenses of $9,250. The percent of households over twenty-five years old with a college degree was 23.38 in 2000. This, as in Santa Barbara, illustrates a high level of social and financial capital in the area, and according to interviewees, is tied to both cities' quality as a destination for retirees. These are the contexts of the grassroots/grasstops tensions that I examine in this chapter.

Eleven California interviewees and thirteen Idaho interviewees had worked or currently worked as paid staff for a nonprofit or political organization. In Idaho, these organizations included the Idaho Sierra Club, Idaho Conservation League, Snake River Alliance, Conservation Voters for Idaho, Advocates for the West, Lake Pend Oreille Waterkeeper, Model Forest Policy Program, Friends of the Clearwater, and Friends of the Earth. In California, these organizations included the Environmental Defense Center, Community Environmental Council, Democratic Party of Santa Barbara, California Student Sustainability Coalition, Central Coast United for a Sustainable Economy (CAUSE), Center on Race Poverty and the Environment, and Food and Water Watch.

In both settings, I interviewed current, past, and future employees of what community members considered the "big greens" or "old guard enviros"—what I call grasstops organizations throughout this chapter. The vast majority of interviewees, however, were volunteer members of grassroots organizations. Though these organizations were sometimes nonprofits, I differentiate them from grasstops organizations because they tend to be younger, typically do not have dues-paying members, and have fluid leadership structures. For these reasons, they are less concerned, or not at all concerned, with convincing boards, raising money, and following mission statements—all aspects of nonprofits that interviewees critiqued. Many of these interviewees were involved in multiple organizations and sometimes sat on the boards of the big green groups.

Like Cable, Mix, and Hastings (2005), I highlight the perspectives of grassroots activists in grassroots/grasstops coalitions. Unlike these authors, I hold out hope for the possibility of collaboration between what they call "professional environmentalists" and the grassroots. I find that despite their differences and critiques of each other's strategies and tactics, members of each organization type recognize the importance of working together, working across lines of grassroots and grasstops. Cable, Mix, and Hastings (2005) focus on the environmental justice movement, arguing that class and race differences account for tensions between grassroots environmental justice activists and professional environmentalists. While most activists in my research do not fit squarely into what these authors consider the environmental justice movement, or the anti-toxics movement, neither are they "professional environmentalists." That interviewees do not fit into these three categories of the environmental movement that Cable, Mix and Hastings (2005) describe raises questions about how the environmental movement has changed in the last ten years.

My research illustrates how climate change and the increase in and geographical dispersal of extreme energy extraction have altered the terrain of the movement. As Klein (2014) argues, a dispersed mobilization of unlikely collaborators, what she calls "blockadia," has emerged. Therefore, the stark inequalities in race and class that Cable, Mix, and Hastings (2005) identify as blocks to grassroots and big green collaboration are changing in some areas. Sacrifice zones are more likely than ever to be located near educated, middle-class, and white communities—the homes of many environmental professionals. For the most part, my research analyzes grassroots/grasstops collaborations among people with similar race and class locations. Race and class therefore inform, but do not explain, the tensions they experience.[2] Even without racial and class divides, there can be tensions between the grassroots and grasstops.

The analysis in this chapter contains two sections. The first outlines the core divide among these sectors of the movement—the value they place on pragmatism. The second describes ways people try to overcome this divide.

PRAGMATISM

Like so many ideas in the realm of resistance to extreme energy extraction that I write about, such as political labels (see chapters 3 and 4) and success (see later in this chapter), activists contest the meaning of pragmatism.

Organizational Form

The organizational form within which activists work informs their understanding of what is practical, and what is not. Organizations with paid staff are constrained by the need to fundraise. This need puts them in direct competition with other organizations in the community. To differentiate themselves in the hopes of winning over supporters, they create narrow mission statements and then, because of the organizational weight that depends on securing funding and the approval of the board, feel compelled to follow those missions. Therefore, these grasstops organizations operate in silos. Becca Claassen had experience with a variety of organizational forms and coalitions. She cofounded 350 Santa Barbara as a horizontal grassroots group, then partnered with Santa Barbara's old guard environmental organizations and the Santa Barbara Democratic Party, and finally, worked for Food and Water Watch, a national organization. She summarized the tensions that result from organizational forms:

There's something called the nonprofit industrial complex [. . .] this phenomenon where you have nonprofits who are obligated to do fundraising in order
to fund their work. [. . .] To prove to donors that they are being successful,
they often have to have very incremental small goals, narrow missions—and
[they] don't like to stray from that. [. . .] It creates this disadvantage for
people who want to create change. A, [nonprofits] can't get political unless
they have a C4 component, which very few nonprofits do.[3] And then B, it
kind of puts them at odds with other organizations, like highly competitive.

In the context of Santa Barbara, where so many nonprofits were com-
peting for the same donor base, it was difficult to break out of these
conventions. In her next sentence, Becca estimated the number of nonprofits in Santa Barbara at twelve hundred. While inflated, this number
demonstrates how salient competition is in Becca's assessment of the
organizing climate.

According to interviewee Ben Otto, energy associate at the Idaho
Conservation League (ICL), a challenge for this member- and dues-
dependent organizational form is that fewer and fewer young adults are
"joining" organizations. Ben saw this as one explanation for the older
demographics of ICL's members. As Ben explained,

So the classic environmental group membership is people fifty, sixty, seventy
years old. [. . .] They kind of let us do our own thing and if they like one
issue, they are gonna become members of the whole organization and support everything. And they will [. . .] write a letter to the editor, that kind of
thing. And that's good [. . .] but how do we attract younger folks who [. . .]
are not joiners of a group writ large? In my impression, they tend to be more
motivated by individual issues. [. . .] They're not going to be a member of the
organization giving us thirty bucks a month, but if we find an issue they are
really fired up about, then they'll [. . .] get super involved, and be very sophisticated about it, turn out their friends. [. . .] But it's a very different kind of
care and feeding of your membership depending on where they are in age.

Thus, understanding how to attract people through issues, rather than
through a mission or member identity, is particularly important if organizations are to gain long-term support from younger generations. Grassroots activist Borg Hendrickson echoed Ben's assessment. She thought
motivating people by focusing on the issues, rather than on organizations,
is key for effective organizing. To accomplish her goals, she had formed a
loose network, rather than an organization.

Being a paid staff member like Ben, however, can create distance
between volunteers and oneself, challenging staff's ability to connect
with volunteers. Becca Claassen, who had always motivated people by

being a model organizer who did everything and said yes to everything, found that in her new role, when I interviewed her, as paid staff who supervises volunteers, she could no longer say yes to everything. This limited her ability to motivate others through example. Borg Hendrickson's reflections on effective communications supported this point about distance. Borg felt that formal organizations—she organized as an individual member of a network involving big greens and grassroots groups—were more constrained in their communication. They have "this little format to it [and] have to have board approval for everything." As a grassroots activist, she was personal in her communication to supporters, and in return, people took the initiative to share important information with her. "My communications with people are me talking to you, so I think, that's part of why people felt so free to just 'Borg! look at this!'—felt so free to just contact us one-on-one, with a quick phone call or email. I think because there's probably something about my emails that isn't quite the same as an organization's emails."

There was also a sense that the staff role could detract from the joy of organizing, both for the staff person and volunteers. Emily Williams worked for the California Student Sustainability Coalition but decided that she ultimately wanted to do organizing on the side, rather than as a paid position. She did not like the power dynamics of paid organizing. Echoing Becca Claassen, she disliked that there was the "nonprofit industrial complex," where campaigns could be used as a selling point to funders. She also felt there was a "weird" power dynamic of being one of the few paid people working alongside volunteers. Ultimately, she concluded that there was a place for paid organizers, but that they should be "supporting or maintaining, but perhaps not leading, the fight." Kyle Fischler thought that making organizing a career would contribute to loss of "some of the draw and enjoyment you get out of it."

As a volunteer, Miranda O'Mahoney highlighted how working with paid organizers could detract from the joy of organizing because volunteers could become less invested and could be assigned tasks that did not inspire them. After being an activist in multiple student groups, Miranda had worked long hours as a volunteer on the Measure P campaign under supervision by paid staff. She felt tired after the Measure P campaign, where she had little responsibility beyond assigned tasks. She explained, "I don't think that we really reached them [the majority of Santa Barbara residents] very well, um, I don't know how we would have either, but *that's up to the paid organizers to think about* [small laugh]" (my emphasis). In this statement, she illustrates how she gave up some ownership of the campaign since there were paid organizers. As she went on to describe going door to door, it was clear that her description lacked luster when compared to her descriptions of the creative events she participated in with student groups. I asked Miranda if she had noticed a difference in organizing experiences with paid campaigners and grassroots organizers and explained how in the earlier grassroots-led phase of the Measure P campaign, which Miranda had not participated in, the campaign was more horizontal. "Before we [the Measure P campaign] had paid organizers," I said, "we [students] all felt important and valued, like we could be leaders in however capacity we wanted to be." Miranda "did not feel like that." I described my own alienation from the Measure P campaign after it became hierarchical, and how I preferred grassroots volunteering because "when everybody's a volunteer, everyone gives what they can, and that's really awesome that they are giving anything." This resonated with Miranda, who replied, "That's kind of what I was trying to get at, like with everybody giving what they can [. . .] *if you are doing what you are best at, and you are giving that, and you love it, then it's like the best situation*" (my emphasis). Hierarchical campaigns with preestablished strategies can forestall possibilities for volunteers doing what they love.

Being a paid staff member also comes with a level of comfort that some grassroots activists yearn for and others critique for affecting one's capacity to be radical. On the comfort side, interviewees mentioned that they could not justify spending too much time organizing, "since it's not currently a paid gig," as Kyle Fischler explained. Helen Yost, who lived in a van full of Wild Idaho Rising Tide campaign materials during this research, yet had pursued a PhD, illustrated both views. She had wanted to earn a living from being an organizer. Yet, since Friends of the Clearwater fired her, she had dedicated herself to the grassroots group Wild Idaho Rising Tide, whose supporters occasionally donated enough money to pay for an office. Helen had become very critical of comfort:

> You sort of just got to throw reason out the window because reason has more or less been co-opted by government and industry, and you just sort of have to follow your heart and say, "Well [. . .] I don't know how I'm going to pay the rent. [. . .] I don't know if I will ever be able to talk people into doing hard-core activism." [. . .] You just say, "Fuck it, I'm going to do this, and I don't really care about anything beyond that"—which is also very freeing because I just sort of say fuck money, fuck a home, fuck prestige, what people think about me, any kind of normalcy.

Helen rejected most elements of comfort that comprise the average American and environmental professional lifestyle. As she explained, "I like to call America the place where you can have whatever you want but nothing that you need because everything you need is being destroyed by what everybody wants." She was highly suspicious of the reasoning behind grasstop tactics and wished she could inspire more people to engage in direct action. As she said, why do anything reasonable if reason was co-opted by government and industry?

This rejection of comfort resonated with Cass Davis's journey as an activist. Unlike Helen, Cass came from a working-class family in an Idaho sacrifice zone—the Silver Valley. In the 1970s in the Silver Valley, Bunker Hill lead smelter and zinc plant decided to pay fines rather than fix their pollution control system, which had been damaged in a fire. In the years following the fire, local children's lead levels averaged fifty micrograms of lead per deciliter of blood; the Center for Disease Control considers five micrograms high enough to warrant concern (Christian 2016). Cass struggled with disabilities that he linked to lead poisoning and so did not have a steady job. He connected the comfort of having a paid environmental job to tame tactical choices: "I would like to say, if I had been able to [. . .] finish college and get multiple degrees, that I would still be crazy enough to break laws in order to disrupt what was happening [climate change], but maybe I wouldn't. Who knows? Maybe if I was comfortable making $40,000, $50,000 a year, doing great things for environmental groups [. . .] I would be happy as hell and kind of trying to hide from that whole extinction trip and just being comfortable." Helen and Cass were the two interviewees who most embodied an anti-capitalist lifestyle. Making money from being an activist, then, was antithetical to their philosophies on life. While their contrast to someone who works as a "professional environmentalist" is extreme, it highlights some of the central divergences between professional and grassroots activists. The biggest one is trust in the status quo, which determines what activists view as pragmatic. Scholars have found more distrust of government and the corporate class among anti-toxics activists—typically grassroots—than among professional environmentalists (Brown and Mikkelsen 1990; Cable and Cable 1995; Krauss 1989).

Grasstops activists, who make a living from organizing, are more attached to the status quo because it gives them stability. They are wedded to a form of democracy that relies on certain types of policy models and regulatory frames that allow some change, but do not ultimately threaten their jobs or retirement. This is not to say that grassroots activists are not also concerned with stability and restricted by it. Jeannie McHale explained that it was hard to "have the level of activity that begins to match the level of injustice." Unlike Helen, whom Jeannie saw "going full tilt," Jeannie, a professor, felt "held hostage by the very quality of life that the things that we protest make possible." She was not willing to make the sacrifices that Helen made. Unlike grasstops activists, however, navigating status quo political and legal infrastructures as a primary method of enacting environmental change was not the entirety of Jeannie's job. Not being embedded in and dependent on these structures made it easier to imagine alternatives.

One caveat to this is that groups that have been historically marginalized, like the Nez Perce Tribe, depend on resources from the government, from the status quo. Paulette Smith, who works for the Tribe, explained how "it's really hard for the Nez Perce Tribe to stand up and say, 'we oppose this, or we oppose that' without stepping on somebody's toes that has just given us hundreds of thousands of dollars to do this project." Paulette understood what is at stake then, when the Tribe, as an organization, is hesitant to take a stance. Money derived from the very system that has systematically targeted the Nez Perce and their way of life is critical to services that people depend on. This is where Paulette felt that grassroots groups could make a difference: "we [grassroots] can speak for the people that can't, and that's our leaders; that's our lawyers." On the other hand, however, there was a line where standing up for justice was necessary. When some members of the Tribe questioned protesting against the megaloads because they thought it was against the law (see chapter 7 for an account of the protest), Paulette took a stance. "As a Native," Paulette explained, blocking the highway was

> not against the law. What they [oil companies] are doing to us is against the law, and always will be. And that goes down to my core, to the pit of my heart, that's the difference between that sort of law, non-Native law, and our law. [. . .] Treaties were always broken, promises aren't promises, we all know that. And so it's walking a fine line to see who within our community, who within our tribal people will jump on board, because they fear retaliation, that's the reality of it. They think that they will be impacted in a negative way and I don't blame them, but at this point in time in my life, I don't care.

Paulette's excerpt illustrates how injustice is embedded in the systems that grasstops, or tribal governments, work within. Her account points to the need to reveal these injustices and stand against them, while also supporting people and organizations whose social position makes resistance outside of these systems difficult or dangerous.

In sum, among grassroots activists and some professional organizers, there was a sense that formal organizations and nonprofit status created many constraints. It made groups less flexible and their interaction with volunteers more impersonal, hierarchical, and instrumental. Rather than moving from issue to issue and inspiring people along the way, grasstops organizations had to consider how issues fit their mission statements and, in Kelsey's view, had to maintain levels of control over campaigns that were not conducive to creating genuine collaborations. They had to think about building membership *in order* to raise money. For these reasons, in part, Kelsey was leaving her job as executive director of a nonprofit in Idaho. In leaving her organization, she hoped to be more able to let "go of attachments to what [an event] is going to be."

Kelsey's perspective on the importance of letting go of control over events developed while organizing the 2015 Idaho Climate Rally. During this effort, Kelsey realized, "If I want everybody to help, help me with my project, it can't be my project, like they have to want to help me and do it because it feels good for them, not because I am making them." Relinquishing control over an event went against the structure of nonprofits. Kelsey explained, "I don't think that really resonates well for organizations that are managing a huge budget, and managing staff, and fulfilling their mission, and giving grants deliverables. [. . .] There has to be more control over the message and over the people."

Another restricting feature of working for a nonprofit was that it inhibited freedom of expression. Kelsey said that nonprofit staff, and especially executive directors of nonprofits, were always concerned about how others would view their opinion: "You say a sentence and you have to think about, 'What might the legislature say about that? Or, if I was quoted in the paper making that sentence, what would the consequences be? Would I say that to my major donors?' You're just so worried about saying the perfect thing that you hold back and I don't think that's what the world needs right now." Kelsey's critique of grasstops organizations wanting to control everything resonated with grassroots activists. That she had learned, while in the role of nonprofit staff, that this was not an effective way to collaborate with people is a hopeful sign of people's capacity to change their strategies. Yet Kelsey did not feel she could collaborate with people in the way she desired while working for an organization—that is why she was starting a new career.

The organizational form of grasstops makes it difficult for even the most well-intentioned staffers to change the organization's approach.

Strategic Divergence: "Giving Up the Farm"

Understandings of pragmatism are also rooted in groups' strategies, tactics, and motivations—the means they employ and ends they seek. More often than not, the grassroots organizations in my research sought an end to fossil fuel projects. Ending specific projects was part of a larger strategy of keeping fossil fuels in the ground to avert catastrophic climate change. Other goals like accountability and integrity or climate justice, though different, would amount to the same outcome if achieved. Grassroots goals tend to be more focused on global or widespread issues than those of grasstops.

Many of the grasstops organizations focused on the local, and in particular, the local environment. *Protecting* that environment was a common theme, setting these organizations' agendas as defensive, rather than offensive, from the beginning.[4] Environment tended to be interpreted in a restrictive manner, as the nonhuman environment.

A focus on protection and preservation is linked to maintaining the status quo. A failure to define which status quo should be protected can make this mission seem counterintuitive. The Idaho Conservation League's vision, developed in 1973, can be interpreted as being particularly dedicated to the status quo: "Keep Idaho *Idaho*." For whom? Taken out of context of the environmental mission of the organization, this vision is particularly problematic. Idaho in the 1970s, the context of the organization's founding, was home to the white supremacist group Aryan Nation, prison riots over living conditions, and the displacement of thousands of people following the collapse of the Teton Dam. In environmental terms, the '70s were the era of the lead poising of children in the Silver Valley and the continued development of Idaho's nuclear industry, which, since the 1950s, had injected radioactive waste into the Snake River Plain aquifer (USGS 2008). In 1980, the Nez Perce had to protest to demand their tribal treaty rights to fish salmon on Rapid River (see Johnson 2018). Keeping Idaho *Idaho*, then, necessitates clarification about what should and should not stay as part of Idaho. It is an ironic vision for a group, that, like others, grassroots activists often critique for being too invested in the status quo of organizing and regulatory infrastructures.

Grasstops organizations' theories of change—their theories of how the world changes—typically prioritize legal and regulatory change, which tends to be slow and incremental, a politics that grassroots see as inadequate to climate science (see also Rosewarne et al. 2014). Legal and regulatory means of change require that groups have "a seat at the table" with decision makers and industry groups. This, in turn, requires that groups maintain credibility with politicians. Investment in this theory of change informs professional environmentalists' views of outspoken grassroots activists as people "lighting their hair on fire." It can lead professional environmentalists to become over-invested in regulation and lose sight of radical vision. The results can be disastrous. As Montrie (2003) documents in the case of Appalachia, grasstops' devotion to regulation, and their ability to convince the grassroots to back this strategy, rather than one of abolition, led to policy that failed to mitigate the environmental effects of coal strip mining, and paved the way for the mountain-top removal mining technique.

In Santa Barbara, Becca saw regulation leading to a sense of false security among the grasstops and politicians. "It's like, not only does [asking for regulation] make the people who are asking for actual justice seem marginal and extreme, it divides us. It gives elected officials in power [. . .] the easy way out." To underline her point, Becca recounted a story about a woman she met while trying to build organizing relationships in her community—Becca was an adept relational organizer. At a League of Women Voter's meeting where they were discussing the risks of fracking, Becca advocated for a ban, arguing that fracking cannot be done safely. An elderly woman countered her view, saying, "Well, no, if you look at offshore drilling, there were those of us who were saying we needed to push for a ban, you know, forty years ago, but [. . .] we realized that we could just achieve regulations, and now offshore drilling is happening safely all around the world." Becca was horrified that the woman was "patting herself on the back" about this. The BP oil spill had just wreaked havoc on the Gulf Coast, which Becca saw as evidence that offshore drilling could not happen safely. In Becca's view, there was a big difference between achieving regulations and achieving a livable planet.

Furthermore, Becca explained, regulations can give the industry more credibility and legitimacy. Even though Santa Barbara County was "the most regulated on paper" with regard to oil extraction, an oil industry insider had told her that oil companies had gotten away with more "environmental atrocities" in Santa Barbara than in other communities. Becca cautioned: "We constantly have to be thinking, what are we asking for, and is it actually going to get us closer to the ultimate win? Or, is it just going to [. . .] be an incremental step that we can feel good about and actually get the industry [. . .] their foot, further in the door and be a false sense of security for the public?" Forty years ago, in the era of the elderly woman whom Becca encountered, many of the environmental organizations in Santa Barbara formed in response to the 1969 oil spill in the Santa Barbara Channel. Regulation was the eventual outcome of public outcry over the spill. It was not, however, what the public originally wanted, as Molotch (1970) argues. Post-spill, Santa Barbarans had faith in the accountability of the political system, emboldened by elected leaders' proposals for drilling bans. Molotch concluded his analysis at the time with hope that some Santa Barbarans "had come to view power in America more intellectually, more analytically, more sociologically—more radically—than they did before" (142). The bans that Santa Barbarans hoped for never came to fruition. From the perspectives of the more radical Santa Barbara interviewees, the Santa Barbara environmental community instead became regulators—regulators in need of a good shake-up, a topic I discuss in the following pages.

A final key point from Becca's analysis is that of regulatory capture, "the process through which regulated monopolies end up manipulating the state agencies that are supposed to control them" (Dal Bó 2006:203). In her view, establishing regulations goes in the wrong direction because it creates a governmental apparatus with employees that depend on oil production, as she argued was the case in Santa Barbara.

Santa Barbara's Energy division, with twelve staff in 2016, was indeed dependent on industry. The County of Santa Barbara (2013) issued a brochure seeking an Energy and Minerals Division manager explaining that "the division is funded *entirely* through permitting revenues from oil, gas, and mining projects. Increased oil and gas production activity in our County is likely given current trends and *will result in the division expanding* to manage that growth" (my emphasis). Growth of oil is good from an organizational view because it assures the continuation of the division.

Near the time when I interviewed Becca in 2016, Los Angeles was also expanding its oil and gas regulatory apparatus. The city appointed an oil administrator after not having had one for a long time. They created an office of Petroleum and Natural Gas Administration and Safety for the administrator to oversee. While Becca recognized that Los Angeles "hadn't had the oversight they needed," she felt like installing an administrator, rather than, for example, seeking to transition to renewable energy, was a "step in the wrong direction." It was creating more social and political infrastructure to support oil in a context where activists advocated dismantling fossil fuel infrastructure.

Idaho's Oil and Gas Conservation Commission is another example of regulatory capture. One of the Commission's goals for 2017–2018 was to develop policies that "foster, encourage and promote the development, production and conservation of oil and gas resources" (Idaho Oil and Gas Conservation Commission 2016). The agency's definition of "conservation" has been used to justify the need for unitization of land. Unitization, or "forced pooling," works to enlarge the area of gas that is extracted "to prevent waste." Forced pooling is a method whereby an oil and gas company can ask a state to force a certain percentage of mineral rights owners in a certain area or "pool" to lease their mineral rights, against their will, if a certain percentage of other mineral rights owners have agreed. In Idaho, this threshold for integration is 55 percent. This means owners of 45 percent of minerals have no choice over whether they want extraction to occur beneath their homes. People who do not own minerals have no say at any point. The bottom line, which resonates with the general grassroots distrust of government, is that regulators who allow policies like forced pooling are not people that grassroots activists want to work with.

This distrust of government directly contrasts with most nonprofit approaches. Linda Krop, chief counsel for the Environmental Defense Center in Santa Barbara, typically met with agency staff and counsel before beginning ballot measure campaigns. While agency staff cannot take a side on a measure, Linda explained that they can identify elements of the measure that they find unclear or of concern, which allows ballot measure proponents to address these before beginning their campaign. To Linda and other activists' dismay, the County played a large role in damaging the Measure P campaign when it published a report by County staff with incorrect information. Linda explained: "It really hurt us, because it wasn't just the oil companies against the environmentalists or the community. It was the oil industry supported by the County disagreeing with us." In Linda's view, better preparation and working with the government might have produced a different outcome. Activists like Becca, however, have less optimistic views about winning regulators' support. As John Foran, who participated as a volunteer in the Measure P effort, said in reply to Linda's assessment: "Maybe the County should have got it right." Indeed, the assumption that inclusion into the legal system will enable justice has been one of the greatest shortcomings of the environmental justice movement and scholarship (Pellow 2016a:384).

These critiques do not mean that professional environmentalists have bad intentions. Critics of grasstops often temper their words with recognition of nonprofits' good work. As grassroots activist Gary Paudler said, "I'm pretty disdainful of those established organizations. They've certainly done some good stuff, but [. . .] ." Many professional environmentalists were inspired to pursue their careers by a desire to make the world a better place—something that also motivates grassroots activists. And some professional environmentalists retain a radical vision, are deeply invested in justice, and wish that grassroots would, to use a grassroots term, "assume good intentions." If grassroots assumed the good intentions of professional environmentalists and their organizations, they might be more open to understanding that they employ particular strategies for a reason.

Ben Otto exemplified this stance. A self-described optimist, Ben understood and was committed to environmental and climate justice, preferred that natural gas extraction did not happen in Idaho, and valued the work of grassroots activists. He wanted them to understand the strategy of professional environmentalists and recognize that it is needed too. Ben explained that his job was about developing relationships with state agencies and utilities and that these relationships informed his communication strategies: "I may be coming across using words that seem more reasonable, or that I'm giving up the farm, or [. . .] conceding a lot of stuff." He wanted the grassroots to know, however, that he was not "selling out" or "trying to get a job at the utility." Rather, he was engaged in a strategy that he felt contributed to grassroots' capacity to achieve their own goals. "Even if your partner [. . .] seems to be playing the inside game, it's for a reason, and there's a strategy, and you need to have both of the things happening in order to get these wins. I have huge respect for the grassroots that show up and say we need to close every single coal plant this year. That's really important, but that's not really going to happen." Ben's plea to the grassroots aligns with calls by activists to employ diverse strategies and tactics—to "use all the tools in the toolbox," as Brett Haverstick advised. Embracing diverse tactics has helped social movement coalitions succeed (Beamish and Luebbers 2009). In reference to his efforts to close coal plants, Ben explained: "That's something that a diverse conservation community needs to get better at understanding—that shared goal [of closing a coal plant] and being comfortable with slightly different tactics and strategies to get there. Because that's how we'll be successful, is the broad coalition." That Ben felt he needed to explain this to the grassroots, however, reveals a disconnect in communication between the two wings of the movement. Rather than working together behind the scenes and then deploying various strategies or marshalling "dual power," which the climate justice movement has done in settings like Richmond, California (see LeQuesne 2019), Ben's account assumes these strategies are developed and deployed without much conversation across organizational types.

Ben's perspective ultimately comes back to working to achieve what is practical. As he said, the hardline grassroots call to keep coal in the ground is "not really going to happen." While he acknowledged the danger of working with agencies and how this collaboration might be interpreted as "conceding" or "selling out," his work to build relationships with them depended on the assumption that they can be moved to work in the interest of the people. Many grassroots activists have lost all hope in state agencies. A barrier is created when both sides feel the other's whole approach is "not going to happen," whether because agencies are too corrupt or because demands are too radical. Having conversations and tempering language to encourage, rather than discount, different visions and demands, or to understand clearly how those demands come together, is key to avoiding the misinterpretations of strategy that Ben described.

Tactical Divergence

These divergent approaches or strategies have divergent tactics. In simple terms, the more reformist approach asks nicely and works to persuade, whereas the more radical approach attempts to make demands. Both camps are playing a game. They each have strategies and different interpretations of how to win.

The professional approach of working with decision makers requires having a seat at the table and offering plans that will be accepted. In contrast, many grassroots activists' actions start with the assumption that oil and gas companies are fundamentally nefarious and have the ear of regulators. Alma Hasse, for instance, often worked to understand and counter what she called "the oil and gas playbook." Part of that playbook is to make activists seem extreme, to appear as people with their hair on fire to regulators, the media, and especially to fellow community members and organizations. The playbook utilizes a strategy of divide and conquer, which is common in extraction zones (see Bell 2016).

The very policies surrounding extraction also incite divisions. Courtney Washburn, executive director of Conservation Voters for Idaho, lamented how regulations for natural gas were developing in Idaho: "It is setting neighbors up to fight with neighbors for years to come because of personal property rights and planning and zoning issues that just went largely unaddressed." Having strong relationships between and among community members is one way to counter this—something that, as discussed previously, can be challenged by the distance between paid and volunteer activists.

Helen Yost worked to counter industry's playbook by ensuring activism was unpredictable and caught the fossil fuel industry by surprise. She saw her strategy as playing "hard core psychological warfare," explaining: "The only way you are going to get them [Big Oil] if you are a rebel, is using all kinds of unknown—to the standard procedure or to the people in control—asymmetrical warfare techniques. You just make it up as you go, you get unpredictable, you get people popping up in God knows where out of the bushes or wherever and it scares the crap out of them [Big Oil], and that's as it should be." Activists in Helen's group, Wild Idaho Rising Tide, and in the network Fighting Goliath "popped up" by monitoring megaloads all over Idaho and Montana. Activists would follow the loads, video their movements, and record their traffic delays. Based on the success of the megaload campaign in Idaho and the rapid success of the signature-gathering campaign for Measure P in Santa Barbara (see chapter 7), an element of surprise serves activists well. Borg Hendrickson described how she and her husband, Lin—core members of the Fighting Goliath network that defeated the megaloads— caught Big Oil by surprise in their rural Idaho town: "A lot of what one has to do along the way is be novel and creative and imaginative. [. . .] You have to do what they [Big Oil] don't expect." Lin and Borg, therefore, turned out about one-sixth of their small town—100 out of 650 people—to an informational session for Exxon/Imperial Oil, hosted by the Idaho Transportation Department. Ahead of time, they found out what messages Exxon/Imperial was going to communicate. "When we got to the meeting," Borg recounted, "We had the truth about every single one of them [the oil company's points], something they didn't expect. [. . .] And they had dubbed this as an information session, not an exchange—we brought our own microphone and speaker [. . .] that, they didn't expect. [. . .] There were a lot of things that we did that were really novel and creative, so that they had no idea how we were gonna hit them at any point in time. And we always hit them with the truth."

Activists and organizations have little ability to create surprises during regulatory procedures. Regulators, however, often did create surprises. During my fieldwork in southwest Idaho, the Idaho Department of Lands scheduled surprise meetings on very short notice multiple times. This made activism feel like running to put out fires. Sherry Gordon explained, "We just keep [. . .] taking the opportunities [. . .] that are thrown at us [laughs]. I don't mean going out and getting opportunities, like selecting one [laughs], no, you just grapple with the one that got thrown at you." The megaloads, however, are one example of a surprise regulatory approach by activists, who, according to Natalie Havlina, put together a successful legal challenge in record time. Thus, activists had mixed success getting ahead of the next step in the fossil fuel industry's playbook.

Grassroots activists extended the playbook analogy to nonprofits as well. Gary Macfarlane worked for Friends of the Clearwater, a group that had only two staff at the time of our interview. Because of this, it shared more similarities with grassroots organizations than large nonprofits. Gary described Friends of the Clearwater as more "feisty" than most big green groups. With the big greens, Gary said, "there's an effort to cut back room deals with supposedly all interests represented at the table. [. . .] I think it's a very anti-democratic effort wrapped up in a nice package called collaboration and working together. [. . .] We're not part of that process, and we have been very critical of that."

Being fed up with the standard procedures of this playbook, what Gary Paudler called "environmental passivism," inspired activists to form new groups. Groups like 350 Santa Barbara and 350 Idaho attracted activists because they filled a perceived need for a different kind of organization. Max Golding, 350 Santa Barbara's cofounder, began his climate justice activism when he bought himself a birthday ticket to the Tar Sands Blockade action against the Keystone XL Pipeline in Houston, Texas. While there, he attended trainings on direct action. This experience was the impetus for Max to cofound 350 Santa Barbara in early 2013. He recounted, "Suddenly the gears are ticking in my head, and I was like this is stuff that I could take back to Santa Barbara. There isn't really anything fiery in Santa Barbara. There's just this nonprofit, you know, water bottle and shit kind of mentality." Recycling or not using plastic water bottles is something grassroots activists saw as a first step—personal environmental behavior. They considered it a tame tactic, and most thought that it was not "doing us any good"—that people should focus on political and system change. Thus, to use it to describe nonprofits reveals Max's disagreement with their tactics and strategies. Cass Davis would agree with Max. Cass did not recycle, something that irked many of his environmentalist friends. His response was, "If you took the time that you took to make sure your recycling bins were in order and going to the recycling center and used that time to cause conflict against the capitalist structure, you'd be doing a lot more."

Gretchen C., the leader of 350 Idaho, recounted a similar reason for being attracted to 350.org: "[350's] form of activism is a little bit more like civil disobedience, a little bit more in-your-face compared to a lot of the groups in town [Boise] which are more [. . .] on the passive side." Gretchen explained that most groups in Boise, Idaho, were unlikely to protest, because their "whole point" as organizations was to get members, or to convince the local utility, Idaho Power, to "maybe" think about projects like community solar. "Rather than force a demand," Gretchen said, "it's more like: 'Let's make friends with them and try to diplomatically work things out; even if they slam the door in our face, we will still go back with a smile.'" Gretchen hoped that 350 would provide "that side of activism" that the other groups did not have.

Commitment to active protest, creativity, making demands, and public participation defined grassroots tactical decisions. In grassroots organizers' views, these actions were the most effective ways to counteract the collusion of government and the fossil fuel industry—what most identified as the root of the problems they faced.

Motivational Divergence: Fear of Losing

Motivations, alongside strategies and tactics, are a final area of divergence that informs decisions about strategy. A fear of losing undergirds the motivations and cautionary approaches taken by grasstops activists. Fear of loss is also tied to length of involvement in the environmental movement. Interviewees who were most fearful of loss had a history of securing victories and suffering defeat in the organizations with which they worked. This suffering, perhaps, informed their hesitancy to take actions that they thought would result in loss, or even worse, a roll back in previous wins.

Dave Davis, age sixty-seven, was former executive director of the Community Environmental Council and someone who had a long history of involvement in Santa Barbara's civic life. His perspective illuminated this fear of loss position. Dave and other leaders in Santa Barbara's environmental and political community had a number of reservations about grassroots activists' idea to try to qualify an anti-fracking ballot for the mid-term election in 2014. Once grassroots activists decided to gather signatures, in spite of Dave's and others' reservations, Dave knew that his organization "had to be on the right side" of the issue. He had to figure out how to convince his board to support the campaign. The Community Environmental Council board did not want to support the measure. They thought the measure would fail and that the failure would reinforce power in North Santa Barbara County, where the fossil fuel industry is located. They feared that failure would contribute to oil-friendly conservatives taking a majority of seats on the County Board of Supervisors. Ultimately, Dave said, the board thought, "*We have more to lose than we have to gain*" (my emphasis). Dave's work to convince the board resulted in a split vote of 7–5 to support the initiative. A couple of board members became key donors to the campaign.

Despite this, the underlying sense that there was more to lose than gain was a strong current in Dave's account of Measure P, which did lose (see chapter 7). In 2016, when I interviewed him, Dave was wary of negative repercussions that could still be on the horizon and pointed to two outcomes so far. In 2015, a Democrat supported by oil companies challenged incumbent county supervisor Janet Wolf, also a Democrat, known for her environmental leadership. Wolf won re-lection by ten percentage points. But then, according to Dave, Big Oil used what it had learned in that campaign to successfully appoint a pro-oil candidate to the Goleta City Council, making it majority conservative. Each city in the region has a seat on community development organizations, so this added to conservative power at the regional level. Dave feared a takeover of the Board of Supervisors as well, saying that if oil ever controlled the Board of Supervisors, "everything we have done for greenhouse gases to oil restrictions, they are just going to undo those suckers." In Daraka Larimore-Hall's words, Measure P "poked the hornet's nest." Poking this nest made Big Oil more bent on securing its power. And Dave thought losing to Big Oil was more likely after the Measure P failure, which oil companies could point to as evidence of public support for oil in Santa Barbara County.

Dave connected his fears to broader trends as well. He thought environmental protections would be stripped if the White House and Senate ever were controlled by Republicans: "*In the 45 years I've been doing this stuff, you work really hard to get things enacted, but once they go away, they're gone for a generation. [. . .] It is beyond hard to get them back. [. . .]* That's what would keep me up at night—both locally and nationally—if we lose what we got, let alone be able to step out and do something more progressive on fossil fuels" (my emphasis). With a long history of progressive organizing, Dave's fear about losing was grounded in a fear of losing not *abstract* successes—successes that a young person like myself might grow up with as the status quo—but *his* successes. He or others he knew had earned these successes through hard work. Needless to say, Dave likely suffered from lack of sleep in the aftermath of the 2016 election of Donald Trump as president of the United States. President Trump and the Republican majority Congress did exactly what Dave feared; they rolled back numerous environmental regulations and restarted stalled oil pipelines.

This fear of loss by key community members had negative consequences for the psyches of grassroots organizers in Santa Barbara, who began to feel as if they would set back the whole movement if Measure P lost. This worry elevated the stress of the campaign tremendously, exacerbating the hostile environment created by oil companies funneling $6.2 million into the county to oppose the measure. Becca, a grassroots originator of the campaign, explained that her decision to go along with changes in some of the campaign's tactics—changes that disillusioned core grassroots originators of the campaign, Becca and myself included—was motivated, in part, by the fear she felt from big greens. Becca explained, "They just created a lot of fear in me. [. . .] It was like if we lost, it was going to set the whole environmental movement back; it was going to destroy the county, all of these things that aren't necessarily true and depend entirely on whether we decide to keep going or not." Her words illustrate her belief in grassroots power and her dedication to movement building, which can happen regardless of legal or political wins, as in the case of Measure P.

Becca believed that whether a movement is set back by a defeat depends on whether participants continue moving forward—something she and her grassroots colleagues were doing. Looking back, Becca said she would have prepared herself for something she called "concerntrolling," which she described as:

There are people that are with you. [. . .] They say they are with you, and they care about the issue, but they just don't like how you are doing it, or when you are doing it [. . .] and they end up getting really negative. [. . .] I felt it on both sides. It was like the oil industry was really ramping up their campaign against us and then even they—call it the old guard in Santa Barbara—was being pretty antagonistic and far less than supportive of our effort, behind closed doors.

In the early stages of Measure P, seasoned activists in the community "concerntrolled" through efforts to decide how and when the campaign operated. In the later stages, the Democratic Party did control the campaign and displaced horizontal organizing based on a team captain structure with hierarchical organizing (see chapter 7). Concerntrolling and Democratic Party takeover were the opposite of Kelsey's strategy advocating that grasstops groups let go of control to enhance broad-based ownership of the movement.

In some ways, then, Measure P originators' decision to partner with the Democratic Party strategy was a response to fear. It was also a response to fear generated by a particular definition of "success," which focused narrowly on passing the measure. For many grassroots activists, however, the success of any one campaign is not limited to changing laws. Building a movement, inspiring new activists, and sustaining momentum are integral dimensions to grassroots' definitions of success. As Becca said, success depends on "whether we decide to keep going or not." She expanded on this point:

> I was aware that there was a possibility that we would lose, but I thought that that was just the starting place. I didn't think that a failure on the ballot was going to be a failure for everyone. I, and a lot of the people I was talking to, felt like we needed to start where we start, and that way, we identify other people who care about this—identify opinion shapers who were willing to go public, spokespeople who were facing the problem directly in their backyards—and all that has happened. I mean, we've got a huge list of people and connections all over the county.

The Measure P campaign inspired many people and built a movement, but it sacrificed some of its focus on this goal to try to secure a policy win, and in doing so, may have made it more difficult to secure both. As Becca explained, if the grassroots had maintained their horizontal strategy, "we would have done a lot better than we did, even if we didn't win." For Becca, better refers to inspiring people to join the movement—"building power."

Since 2014, however, there are many signs that Measure P did not have the negative ramifications that the big greens feared. On the local level, it likely enhanced support and political will for progressive policies. In 2015 and 2016, the Santa Barbara County Board of Supervisors approved the state's most stringent carbon emissions levels for oil projects, funded a feasibility study for community choice aggregation, and rejected an oil project with stricter environmental protections than a similar project the same board approved just three years before. Despite challenge by an oil-supported candidate, former environmental attorney Joan Hartmann won the contentious Third District supervisor seat in 2016, maintaining a progressive majority. On a broad level, the anti-fracking movement has grown stronger since Measure P's defeat. Monterey County became the first oil-producing county in California to ban fracking. Its measure also banned *all* future conventional drilling. Just a month after Measure P's failure, New York state banned fracking in 2014. These wins illustrate the grassroots view that energy mobilized and inspired through community and movement building does not dissipate with a defeat. It remains, grows, and spreads as people travel and communicate across blockadia. Tsing (2004) calls these traveling sets of tactics, plans, and inspirations "activist packages." Helen Yost described this phenomenon in relation to the megaload struggle, which went on for several years before it succeeded in stopping the megaloads: "We were just little guys in the streets, and you know what comes next? There are a lot of little towns and other little guys watching this whole thing happen, and they are going to rise up too because they were pretty freaking inspired."

Grassroots activists then, viewed losing in a different way. Many saw the struggle, rather than the loss, as the defining feature of their work. Their decision to act and risk loss was informed by a commitment to doing what was right, not necessarily what was politically or legislatively feasible. In Natalie Havlina's words, "you don't always win, but you fight because it is the right thing to do." Cass Davis, who was particularly bleak about human capacity to survive climate crisis, shared a similar perspective: "I know the battle in sight is not winnable, I've always kept in mind: 'You don't fight fascism to win, you fight it because it is fascism.' There is no winning, just anytime fascism raises its head up, it's repressing, it's horrible, and so you fight against it."[5] These perspectives are based on a more radical understanding of success than most nonprofit definitions.

In contrast, because environmental groups have very little power compared to the gas company in Idaho, Ben Otto of Idaho Conservation League advised that environmental groups learn "how to lose well [. . .] the same idea as kind of losing the battle and winning the war." He felt that even if natural gas production occurred—a loss for environmentalists—ensuring that it was done as safely as possible would ultimately be a win. Many of the regulations that Ben envisioned as part of a "win," however, never materialized. Even if they had materialized, grassroots activists would likely have viewed them as incremental steps, nowhere near the level of change needed to avert climate catastrophe.

Before moving on to grassroots ideas for bridging this gulf between sectors of the movement, I highlight a point of contrast in grasstops' relationship to losing. Ben's concern with losing "gracefully" was different than Dave's fear of losing it all. Context informs this contrast. In Idaho, groups like Ben's had long been working in a hostile political climate with no hope of changing that climate. Idaho has had a Republican legislature and governor since 1995. In December 2016, of the 110 people in the legislature, only 17 were Democrats. Because of this, it is understandable that Idaho grasstops worked to maintain their seat at the table and play the game: it was the only game available for working toward the types of change and successes that their organizations prioritize.

Groups in Santa Barbara, on the other hand, had a long history of securing regulations. While the state and community provided a far more favorable political climate for environmentalists than Idaho, there were periodic wins for conservatives in the state and county—just enough to add a sense of precarity to the liberal status quo. Wins that occurred under favorable circumstances could change if the political context changed. In contrast, any win secured in Idaho was secured within a hostile environment. Policies that pass tend to be palatable to conservative Idaho politicians and therefore are unlikely to be jeopardized by future conservative administrations. These two contexts contributed to whether activists feared losing or planned how to lose gracefully.

CREATING A UNIFIED FRONT

The divergences in understandings of pragmatism, evident in divergent strategies, tactics, and motivations, produced tensions because most interviewees wanted to work together. They recognized the importance of a unified front—of having all progressive groups put aside their disagreements to work together. As Helen Yost explained, "there aren't enough hours in the day to talk about what is wrong with the other guys [other organizations] and still do what's right with yours." *How* to achieve unity, however, is an open question. Understanding where each individual or organization is coming from—how their organizational form and experiences shape their decisions—is a place to start. First, however, in line with the thesis of this book, people have to be able to come together and talk, which depends on relationships of trust. Interviewees highlighted specific practices that make this difficult and ways to change these practices.

Don't Stifle New Ideas

Grassroots activists felt stifled by grasstops. There was a sense that these organizations wanted to control the situation and were against "upstarts," new ideas, and different tactics—particularly when these came from young people. In response, grassroots activists advocated openness to new suggestions, new energy, and new approaches.

Kelsey, whose experience planning the first Idaho Climate Rally helped her realize that effective collaboration required nonprofit organizers let go of control over campaigns, suggested that nonprofits work to support new groups and new ideas, rather than shut them down. There needed to be a balance between established organizations' experiences and allies on one hand and new groups' enthusiasm and potential for success on the other.

When activists first came together to plan the 2015 Idaho Climate Rally, big green groups in Boise thought that the other groups would "jump on board" with their plan for getting Idaho to close coal plants. Grassroots activists reacted with questions about how the big greens developed the idea, and their own proposals. As Kelsey explained, the grassroots' view was, "'We are not going to sign on to a plan that we didn't help develop.' So it wasn't even that they thought our plan was bad, it was more they didn't help make it, and they wanted more information before agreeing that it was a good plan. And that's legit," Kelsey explained. For Kelsey, this disagreement revealed a divide between the more established groups, who "are more skeptical, maybe more jaded," and newly formed groups. Some of the new groups wanted to push the city of Boise to have 100 percent renewable energy or to cut off its relationship with Idaho Power and have its own power generation. Older groups, who had worked with the city for decades, were skeptical that this would be successful and also wanted to preserve the relatively good relationship they had with the city.

Kelsey wished that the big greens would recognize that "we don't want to stamp out the enthusiasm of somebody who wants to try again. Like we are quick to say, 'Oh that won't work and here's why, or we have already tried that, and they said no.' Why would we want to stop somebody else from trying again? Like maybe they will be able to do it." For her, an important question for the nonprofit community was learning how to acknowledge different approaches and support them, no matter what similar campaigns had yielded in the past:

Everyone has their own opinion based on what they learned and so how can we teach each other what we learned and not dampen anybody's enthusiasm? As [someone] younger [age 34 . . .] than a lot of environmental activists, I hate when like my idea is stamped out. [. . .] I'm kind of like, "Screw you; you couldn't make it work, that doesn't mean that I can't." [. . .] So we wanted to make sure that we weren't doing that—[so] a new group forms, and their idea is to get Boise off Idaho Power; like, go for it. Let's help them figure out how that might work as opposed to telling them, "Oh you should stop; that's a bad idea."

This type of mentality—open to new methods, supportive of new enthusiasm, and optimistic—was the mentality that grassroots interviewees longed to see among grasstops groups. It is markedly different from how grasstops responded to the grassroots organizers of Measure P.

The big green groups working on the Idaho Climate Rally recognized that, at the beginning stage in their relationship with the grassroots groups, having a unified policy position was not necessary. It also was not a prerequisite for working together. "We decided," Kelsey said, "that instead of using our planning meetings for us to persuade them that our plan was right, we would use our planning meetings to plan this event where we all would be able to say what we wanted and then we would continue to work together in the future." These groups agreed to disagree about their long-term strategies in the interest of working together on a shared campaign in which they could learn more about each other, their ideas, and hopefully build relationships that would enable future collaboration. This is a good example of what activists view as an effective way to support each other across different organizational forms. It is the classic method for justice-oriented community organizing—where people come together and develop plans and dreams together, in collaboration.

Left Flank: From Margin to Center

Working together also requires respecting, valuing, and supporting others.[6] Unfortunately, the divergence in visions of what success means leads both sides to discount the other's approach. While I, and most other grassroots activists, are sympathetic to the draw people feel to work on solutions that seem possible in the current political climate, climate science demonstrates that the solutions developed in this realm so far are too little too late (Holdren 2014; IPCC 2014)—in this sense, they are not pragmatic (Rosewarne et al. 2014). Having any chance of staying *around* two degrees Celsius global average temperature rise requires radical change. Tim DeChristopher's opening quote to this chapter summarizes grassroots' sentiments well. DeChristopher, a grassroots activist, spent twenty-one months in jail for outbidding oil companies in an oil and gas auction in 2008. He does not believe the solutions advocated by the NGO-led climate movement will work, explaining "There are very few things that make me more hopeless than a movement based on useful fictions" (Tim DeChristopher quoted in Stephenson 2015:190). The primary fiction that big greens operate on is a belief in working with power holders to regulate environmental damage, in essence, to establish tolerance levels. Grassroots activists, on the other hand, demanded no more damage.

None, Not How Much

Jane Fritz and Becca, activists from different eras and in different states, thought it was absurd that the respected and well-resourced nonprofit groups in their communities were negotiating over *how much* pollution would be allowed. Sociologist Ulrich Beck points out that this conversation about risk thresholds is not about protection but, rather, an acquiescence to a level of acceptable poisoning (Beck 1992:65). Experts have the privilege of setting that level, not people who are directly affected by the poison. Grassroots activists shared this view, seeing this situation as fundamentally unjust for people and the environment. Jane and Becca, whose stories I recount in the following paragraphs, held fast to their stance that there should be *no* risk. In their view, environmental groups should not be in the business of deciding *how much* risk is acceptable—this was antithetical to their purpose.

For Becca, the issue was mitigation of oil extraction emissions for the Santa Maria Energy Project. Becca's group, 350 Santa Barbara, felt that their push to reject the project completely, when other environmental groups were pushing for regulation, contributed to stricter emissions policies. That big greens were only seeking mitigation, "meaning oil companies would have to pay some cents for every ton of CO_2 beyond ten thousand tons per year," was horrifying to Becca.

For Jane, the issue was herbicides in Lake Pend Oreille, a place to which she was deeply connected. She recalled:

> When you have four of the top environmentalists in town sitting down and deciding how many thousand acres they are going to treat with 2,4-D, I am sitting there going, "Are you guys crazy? You don't give them a number!" [. . .] And the county was so clever. [. . .] : "Oh well we'll form a task force

and bring them [the environmentalists] to the table." Well, when they are at the table, they want to cooperate, they don't want conflict; most people don't want conflict. So as a result, they [the county] got everything they wanted [laughs]. And it was the environmentalists who gave it to 'em. And I just said, "That's it, I'm out of here, I can't deal with you people!"[7]

Jane's feelings were: "You are poisoning my mother!" Jane had been an activist in Sandpoint, Idaho, since the early 1980s. She had seen the community go from one where 350 people would show up to protest at public meetings to one where environmental groups held happy hours at the local brewery for their members. She did not necessarily think that hosting meetings at a brewery was bad, but she felt like the movement was weaker and more complacent than in the past. She lamented the fact that the younger folks running the movement had little interest in learning from older activists like her.

Becca, Jane, and the groups they worked with represented what grasstops thought of as the "left flank." There was an understanding on both sides that radical demands made the other demands look more reasonable. The left flank shifts the window of polite conversation a bit to the left, enabling change on some levels. In the fossil fuel divestment movement, for example, Bergman (2018) finds a "radical flank effect" where activists framing the fossil fuel industry as a public enemy enabled shifts in mainstream discourse around stranded assets and carbon bubbles (the idea that investments in carbon will become stranded, or unprofitable, as policies to alleviate climate crisis are enacted). While the mainstream can achieve change by positioning itself in contrast to the radical or left flank, from the perspective of grassroots activists, this dynamic is a problem. The problem with the left flank is that it is left. In other words, it is a marginalized fringe position that gives centrists the ability to say that it is unrealistic, radical, and idealistic, and that their own position—which can often involve negotiating over an "acceptable level of poisoning"—looks like the realistic, reasonable, and pragmatic action. Furthermore, mainstream notions of what is reasonable change over time. Becca explained this in reference to the abolition of slavery:

We have to build power, and you don't do it by bullying people or advocating for incremental steps that don't inspire masses of people. [. . .] Did the antislavery movement ask for regulations on slavery? Sure, there were some people who were probably pushing for regulations, but no, the powerful movement came when white people started advocating for abolition of slavery, the outright end of slavery. [. . .] I don't know why environmentalists don't understand that—that we have to stand up and start asking for what we actually need together.

Objectively, accepting any level of poisoning is not reasonable or pragmatic. It is also unjust. This norm—the rejection of the precautionary principle—is at the root of environmental degradation and climate change. That accepting levels of poison is so routine within the legal and political infrastructures that grasstops target demonstrates just how much control corporations have under capitalism. This compliance is a perfect example of what LeQuesne (2019) calls "petro-hegemony," the fossil fuel industry's power to shape material conditions and ideology through control over culture, the state, and the economy. When I asked one interviewee about their organization's stance on drilling, they wanted to remain anonymous and replied, "Well, we would prefer for them not to drill, but that's not what we are going to say publicly because then you just get chased out of the room." Big green groups' acceptance of negotiating over regulation and poisoning as the norm maintains the gulf between the grasstops and the left flank.

Centering the Left

The left flank is also problematic because it isolates grassroots activists. Becca's feeling of being bullied by nonprofits had made organizing significantly more stressful for her. Helen Yost felt exhausted from being the left flank. She wanted more freedom to "just let things play the way they play and not follow some sort of standard procedure playbook" of big greens. "Even better to not be scorned because you don't follow that rule book," Helen said. Isolating the left flank is part of oil and gas's playbook. Alma explained this connection:

> I've been labeled a fear monger, anti-oil and gas, [. . .] but it seems like the oil and gas industry, they have to do that. They have to apply these labels to you to marginalize and demonize you so that people don't listen to you. [. . .] The message that I hope [. . .] you're getting for your book, Corrie, is that we should always speak the truth, because [. . .] the truth is our biggest friend, you know; you don't have to embellish. [. . .] Truth, the data, is squarely on our side.

By making demands for action that are in line with science seem unreasonable—again, the memorable characterization of grassroots activists as "lighting their hair on fire"—petro-hegemony leads the public and the big greens to discount grassroots demands. Grassroots activists wanted respect and consideration, both for their ideas, and as individuals.

Bringing the left flank from the margin to the center means reorienting the status quo to center radical, rather than mainstream, perspectives. As Bella Abzug, founder of Women's Environmental Development Organization, has said, "Women do not want to be mainstreamed into a polluted stream: they want the stream to be clear and healthy" (quoted in Dankelman 2010:15). This type of goal means shifting the entire paradigm to the left. The status quo must change toward justice, rather than radical perspectives being absorbed or accommodated by the status quo. Grassroots activists wanted to change the organizing paradigm, not adapt to a status quo they see as unjust and ineffective.

Also foundational for creating a new center based on perspectives from the margins is recognizing that activists need everyone to change everything—a slogan of the climate justice movement. Big green staff who are not trying to change everything may have more difficulty empathizing with this principle. Yet, nonprofits' dependence on membership and the acknowledgment—by most nonprofit staff—that they are more successful with the grassroots behind them, should attune them to the importance of inclusivity. Likewise, taking the science that informs their understanding of the environment seriously would point to a need for more radical positions.

Grassroots activists can struggle with putting this principle of inclusivity into practice as well. Those who work in geographically or politically isolating environments may recognize the need to have everyone on board, yet be so accustomed to working alone that they alienate others they try to collaborate with. And, as youth activists have demonstrated, working beyond established social networks is a challenge for most activists. Concrete practices that can help everyone feel valued include relational organizing and giving attribution. Activists wanted to have trusting and rewarding relationships with each other and members of different groups. They all recognized that trust was an essential component of working together.

Giving Attribution

Giving proper attribution was a good way to build trust. People always admired an activist who demonstrated humility while crediting others or the group. In Shelley Brock's words, "leav[ing] egos at the door" was important. On the other hand, personal and organizational relationships could deteriorate when one party took credit for collaborative work. The big greens wanted credit for their role in regulatory change. The grassroots wanted credit for turning people out to hearings, which demonstrated the public support necessary for transforming big greens' work into wins. The organizational form of big greens, however, challenges activists' capacity to credit each other and leave their egos behind. Big green staffs' careers are built on their activism. In addition, these groups depend on differentiating themselves from others to securing funding. A key question is, as Ben Otto explained, "How do you tell a story of how different groups did different parts of achieving the same policy outcome and be able to talk about it in a way that is respectful of acknowledging each, to the funders?" Ben recognized this as an area of organizing that "needs a lot of work."

Alongside welcoming new ideas, creating a unified climate justice front requires reorienting actions around a vision that is based in justice. Reshaping actions and visions that everyone wants to be a part of depends on collaboration. Valuing each other through giving attribution, relational organizing, and joining with, rather than isolating, the left flank are ways activists on both sides of the grassroots/grasstops divide can begin this work.

CONCLUSION

One would think that environmentally focused groups, organizations, and individuals would have an easier time working across lines than the climate skeptics and liberals who come together to fight natural gas in Idaho. Yet, from my analysis of the interviews, this assumption is incorrect. Caring about the environment is merely a different starting point for the process of learning how to work together, like the starting point of being concerned about property, integrity, and accountability, or being a college student learning about climate justice.

During my time organizing and talking with activists, the grassroots/grasstops divide came up again and again. Like the talking across lines happening in southwest Idaho, it was not something I had anticipated. In Santa Barbara, the divide was something I experienced during the Measure P campaign. At first, the campaign was ours—grassroots. It was as horizontal, empowering, and invigorating as 350 Santa Barbara had ever been—everything that drew me to join the group in 2013. But when our initiative made it onto the ballot, the big greens stepped in and we, the grassroots, felt pushed out of leadership. Our feelings at the time and since then, the perspectives of interviewees, all corroborated the need for a different kind of working together.

In Idaho, I heard about this divide from all sides. I heard it from the radical and often on-her-own Helen Yost, whose dedication to climate justice is less known beyond her community than it should be, largely because of historical challenges working with other groups. I heard it from the tiny nonprofit Friends of the Clearwater, which is more grassroots than grasstops. I heard it from a former attorney for a big green and a former executive director. I heard it from CAIA out on their own in Payette County wondering why no groups in Boise would help them. And finally, I heard it from the big greens themselves. While their views were heterogeneous, most, like Ben Otto, were sincere in their desire to work together to achieve climate justice. They felt some of the same lack of support from grassroots that the grassroots reported about the grasstops.

Bridging this divide would tremendously strengthen the climate justice movement. New ideas, energy, and hope from climate justice activists, Indigenous knowledge, and experience from frontline communities, plus decades of resources and infrastructures of environmental organizations—the environmental movement—would be a formidable force for the fossil fuel industry to contend with. I argue that a commitment to pragmatism—defined as what is possible within capitalism—on the part of big greens is the root of the divide that keeps these wings of the movement separated. Pragmatism is embedded in the organizational form of established and large nonprofits. It constrains their vision of what is possible—shaping their strategies, tactics, and motivations. The status quo that interviewees experience marginalizes anything outside of what current politics and corporate power deem reasonable. It is a status quo that scoffs at radical visions of a just future, that sees emotions of alarm and anger as illogical—as someone lighting their hair on fire. Destabilizing this status quo, especially when all the science demonstrates that radical change and extreme urgency and love for one another is what we desperately need (see Stephenson 2015), should be the work of all people interested in climate justice.

To achieve this goal, people organizing in grasstops or grassroots groups can welcome new ideas and new activists. They can seek to support people with the most radical visions—the visions necessary to inspire real change. Rather than marginalizing those visions and leveraging them to make other changes more politically feasible, activists should go join with the left flank. Together, with steadfast demands for what is needed and a commitment to building relationships of trust that support and congratulate each other, the power of each individual and group will be more coordinated. The movement will have the capacity to, as Helen would say, pop out of the bushes and "scare the crap out of Big Oil." At this stage in the climate crisis, this is more pragmatic than the one-step-at-a-time "make nice with government and industry," to use David Monsees's words, that characterizes much of the work of big greens.

To understand what movement divergences and efforts for creating a unified front look like on the ground, chapter 7 explores two concluded resistance struggles: the fight against the megaloads and the mobilization for Measure P.

CHAPTER 7

Two Tales of Struggle

Coalition Building against Big Oil

With the megaloads [. . .] we built a strong grassroots
movement. We were able to develop relationships that never
existed before and to bring together a dynamic coalition. We
were able to really put forth a strong message, an argument,
that really highlighted why the megaloads, for various
reasons, were wrong—both locally, regionally, and globally.

—Brett Haverstick, Friends of the Clearwater, Moscow, Idaho,
interview

Measure P came blowing out of the ground in December and
January [of 2013–14], and [for] many of us who had been
involved in electoral politics, it was the wrong year, it was
too late to get going. [Originators of the measure] *had not
built any groundwork within the community* in order to be
successful; [. . .] it was like, well why now?

—Dave Davis, Community Environmental Council, Santa Barbara,
California, interview (my emphasis)

By far the biggest difference in California between the
successful [2014 fracking] bans in San Benito and Mendocino
counties and the unsuccessful one in Santa Barbara County
wasn't the funding differences but the *opposing coalition*. [. . .]
Measure P was opposed by a variety of trade and public safety
unions who had been persuaded by oil company deception
into believing that county property tax revenues would be
severely impacted by the ban.

—David Atkins (2014), Measure P campaign manager (my emphasis)

In 2014, campaigns to stop megaloads in Idaho and to ban extreme energy extraction in Santa Barbara County ended.[1] As the preceding quotes demonstrate, participants had very different reflections about how the campaigns went. In Idaho, people who worked to stop the megaloads had a generally positive assessment of how they built a diverse coalition of individuals and organizations that eventually achieved its goal—stopping the transportation of tar sands infrastructure along Idaho highways. In Santa Barbara, participants in the 2014 ballot measure to ban fracking, acidizing, and cyclic steam injection in the sixth-largest oil-producing county in the state expressed just the opposite view. Many, like Dave Atkins, felt that a failure to build coalitions explained the final vote—39 percent in favor of the ban and 61 percent opposed to the ban—though there was much disagreement about which coalitions would have made a difference.

This chapter tells the story of these two struggles to evaluate their strengths and weaknesses and to examine the processes involved in working across lines. Working across lines, in contrast to talking across lines, examines what people *do* together within campaigns and how this affects the outcome. Talking across lines is more concerned with what people say or value as methods for effectively working together. These concepts of course happen simultaneously and can also precede or follow the other. This chapter shows how failure to talk across lines in Santa Barbara made working across lines difficult and, on the flip side, how working across lines in central and northern Idaho helped people get together to continue working across lines.

I find that different levels of commitment to and enthusiasm for diversity—in motivations, tactics, and participants—and horizontal, relationship-focused political cultures, shaped the different outcomes of these campaigns. The chapter reveals the widespread importance of these concepts and practices that are so clearly articulated by youth climate justice activists' values of relational organizing and intersectionality (see chapter 5). It also expands the scope of practices at the heart of how Idaho anti-fracking activists talk across lines by drawing on common values (see chapters 3 and 4). As Brett Haverstick, education and outreach director for Friends of Clearwater, explains in the opening quote, the success of coalitions depends on activists' capacity to advance a visceral message—that what they are fighting is fundamentally wrong.

MEGALOADS: "WEAPONS OF MASS DESTRUCTION"

I think the megaloads are horrible for the environment and most
other things.

—Elementary school student, Palouse Prairie School, Moscow, Idaho

In early 2010, Idaho environmentalists and residents caught wind of
a proposal that alarmed many.[2] ExxonMobil planned to take over
200 "megaloads" along rural Idaho highways from the Port of Lewiston, Idaho, to tar sands mining operations in Alberta, Canada. Exxon
needed a way to get these loads, what it referred to as "modules"—
structures that would become part of Exxon Canadian subsidiary
Imperial Oil's tar sands extraction site infrastructure—manufactured
in South Korea, to the Kearl Oil Sands Project in Alberta. Traveling
through the Canadian Rocky Mountains posed challenges for these
*mega*loads, as many of the routes have tunnels and narrow rock-faced
roadways. Thus, Exxon and its contracted transporter, Mammoet,
identified Highway 12, a winding scenic byway in Idaho's wilderness, as the best route and "a game changer for Alberta's oil sands
developers" (Mammoet Canada Western Ltd. 2009). The modules and
their trailers, each with ninety-six wheels, were shipped into the Port
of Lewiston, Idaho, the farthest inland port from the Pacific Ocean.
The megaload size ranged from 24 feet tall by 24 feet wide by 120 feet
long to 255 feet long and 644,000 pounds; a typical logging truck
weighs 80,000 pounds (Friends of the Clearwater n.d.) (see figure 7).
The average width of Highway 12 is 21 to 24 feet, with little or no
shoulder in many places.

Highway 12 is federally designated as the Northwest Passage Scenic Byway. The same route Lewis and Clark took on their trek to the
Pacific Ocean, the road travels through the Nez Perce Reservation and
along the Clearwater and Lochsa Rivers, both designated wild and scenic rivers by Congress in 1968. With little traffic and no overpasses,
the route attracted Exxon and other oil companies, which negotiated
in 2009 with Idaho's governor and legislators to secure approval to
transport their loads.

Oil companies did not anticipate the opposition marshaled by rural
residents. From 2010 to 2013, delays impacted Imperial Oil, contributing to the company being behind schedule by six months and over
budget by $2 billion (Briggeman 2015). Some of the first street protests
occurred in Montana in 2011, followed by conservation groups and the

FIGURE 7. Megaload wheels. Photo courtesy of Dave King.

Missoula County Commissioners winning a Montana District Court battle that prevented megaloads from traveling on the Montana portion of Highway 12. This prompted Imperial to use a temporary, alternate route, US Highway 95, where it met intense protests in multiple towns, most persistently in Moscow, Idaho. To use this route, which included overpasses, streetlights, and electrical wires, Imperial Oil cut the modules' height in half—something it had previously said was impossible. Cutting the modules in half cost Imperial about $500,000 for each module.

In 2013, a federal judge ruled that the US Forest Service has the authority to regulate megaloads on Highway 12 and imposed a temporary injunction against future transportation of megaloads by Omega Morgan, the company targeted in the case. The injunction, effective until the Forest Service completed a corridor study and consultation with the Nez Perce Tribe, required the Forest Service to issue a highway closure order for a portion of the highway if the State of Idaho issued another permit for an Omega Morgan–hauled megaload. Barred from using Highway 12, Omega Morgan tried other lengthier routes through southern and northern Idaho (see map 2). The expense of these routes, however, eventually made clear the impracticality of the endeavor.

Diversity as Strength

Diversity was a defining feature of the grassroots mobilization against the megaloads, both in terms of motivations for getting involved and the actions protesters took.

Motivations

People opposed the megaloads for their assault on a rural way of life that many residents cherish about Idaho or move there to seek out. Residents feared that Idaho highways would become industrial corridors—all for the benefit of oil companies. They saw no benefits for locals; in fact, they saw many costs. Idaho taxpayers would be left to pay for road repairs on the heels of truly *mega*loads—loads much larger than roads were built to handle. Residents also feared costs to local economies. The Lochsa and Clearwater River corridor is a recreational paradise. Local businesses depend on the area's appeal to tourists. People wondered what would happen if a load fell into the river. How would it be removed? How would it affect the migration of threatened and endangered fish? It would completely change the character of the place. As Gary Macfarlane, ecosystem defense director of Friends of the Clearwater, explained, "Those things are bright and big and huge, and they make a lot of noise." How would a camper like to be woken up in the middle of the night by a megaload? Safety was another concern, as the megaloads blocked both lanes of traffic on a highway that was the only road for people living in the area. What if someone had to go to the hospital?[3]

Another core concern was climate justice. Wild Idaho Rising Tide (WIRT) member Ellen viewed the loads as "weapons of mass destruction" that directly contributed to the oppression of Indigenous peoples, the environment, and the climate. Jeannie McHale, another member of WIRT, explained that letting the megaloads go through Moscow, Idaho, without putting up a fight would make her an accomplice to these injustices. Paulette Smith, a member of the Nez Perce Tribe and Nimiipuu Protecting the Environment (NPE), who was arrested on August 6, 2013, while blockading the megaloads, felt compelled to stand up for her people and in solidarity with her sister in Alberta, who had suffered personal trauma as a result of tar sands development. Some interviewees had traveled to the tar sands region for annual

Indigenous-led Healing Walks and witnessed the destruction. Interviewee and WIRT member Sharon Cousins compared tar sands extraction sites to Mordor, a wasteland and seat of evil in *Lord of the Rings*: "When I saw those trees being taken down like that, [. . .] the black, I have every sympathy with people who call it [Fort McMurray] Fort McMordor. That's what it is—it looks like Sauron's territory." Scholars refer to places like the tar sands region as "sacrifice zones," where the well-being of human and more-than-human communities, who are thought to be disposable and powerless, is sacrificed to benefit privileged communities. In the case of the tar sands region, First Nations' homelands have been desecrated, and cancer rates have risen, all to provide energy to the global market.

Lucinda Simpson, also a member of the Nez Perce Tribe and NPE, described her motivation for resisting megaloads in terms of trying "to stick up for what we need: we are losing a lot of our roots and our fish and the eel." Lucinda explained that many of their traditional foods, and the cultural practices tied to these, are dwindling because of climate change. Like other Indigenous communities around the world, the Nez Perce have little responsibility for climate change, yet because of their connection to the land, they are some of the first to face its consequences. In addition, despite the fact that Nez Perce treaty rights predate the State of Idaho, neither the oil companies nor the state sought approval from the Tribe to transport megaloads through its reservation. The federal government did not seek approval either; before the legal battle ensued, the US Forest Service did not think it had the power to regulate the loads. Thus, the megaloads were also a violation of Indigenous sovereignty (see figure 8).

Many interviewees were deeply concerned about, and motivated by, climate change: "We became potential gatekeepers for practices with apocalyptic consequences," explained Jeannie McHale. Ellen's protest sign displayed a skull and crossbones image that said, "Stop Exxon Genocide." Meryl Kastin described how her young daughter could not understand why people would support the megaloads or tar sands mining, "if we know it isn't a healthy thing." Trying to give her daughter hope fueled Meryl's involvement with the megaload resistance on Highway 12. She explained, "I really wanted her to see that people could actually make a difference, that we could—we could show up somewhere and make our presence known, and that we could write letters, we could call people, that we could make a difference in the world."

FIGURE 8. Nez Perce and allies protest along Highway 12 within the Nez Perce Reservation. They stand next to the Heart of the Monster (grassy knoll on right), where the Nez Perce people come from. Photo courtesy of Friends of the Clearwater.

Tactics and Strategies

With these diverse motivations, megaload challengers employed diverse tactics and strategies to stop the megaloads. They identified this diversity of motivations, tactics, and strategies as the key to their success. In the words of the education and outreach director for Friends of the Clearwater, Brett Haverstick, "It takes a community to stop a bad project. It takes a tremendous team to make a difference [and . . .] you need to use all the tools in the toolbox to bring forth change." The "tools" that activists used fell into three categories: legal, general activism, and protest and blockade.

Legal. Many of Idaho's environmental organizations became aware of the megaloads in April 2010. Knowing that, in Natalie Havlina's words, "Idaho state court is not renowned for being a friendly place for environmentalists," everyone wanted "a federal hook" for a legal challenge. Natalie, an attorney for Advocates for the West who worked on legal challenges to the megaloads, researched the situation from April to August 2010, and, with three residents of the Clearwater Lochsa corridor as clients, she and Laird Lucas, also of Advocates for the West, requested a temporary restraining order in state court, which was granted. In the next year, individuals and the groups Idaho Rivers United, Advocates for the West, and Friends of the Clearwater all participated in various legal actions. In 2011, in federal court, Advocates for the West, on behalf of Idaho Rivers United, challenged the US Forest Service's refusal to take action on the megaloads, and on February 7, 2013, Judge Winmill ruled that the agency does indeed have jurisdiction and a responsibility to protect the values of the wild and scenic river corridor.

General Activism. What Natalie Havlina described as the "general activism piece" of the megaload fight included people going to public meetings, monitoring the loads, and writing letters to the editor. Borg Hendrickson and Lin Laughy, residents along Highway 12, were the focal points of this effort and key players in the legal battle. They formed the network Fighting Goliath: The Rural People of Highway 12, and organized their neighbors. As Borg explained, they worked to be creative and marshal the truth in their favor. In the early days of the megaloads, they turned out over 110 people to a meeting by the Idaho Transportation Department in Kooskia, Idaho, a town with a population of 650 people. Upon hearing that the oil companies were going to have information boards at the meeting, Borg and Lin prepared their own information boards to counter the inaccuracies presented by the oil companies. They brought their own microphone and speaker and turned a one-way information session into a public exchange. Borg developed a large media and contact list to which she sent annotated updates on a regular basis. Communications networks sprung up around the state; a volunteer in Moscow developed a Facebook page to share information.

Along with Fighting Goliath, the grassroots group Wild Idaho Rising Tide (WIRT) and the small nonprofit Friends of the Clearwater (FOC) organized megaload monitoring throughout the region. Members of the groups Great Old Broads for Wilderness, Northern Rockies Earth First!, and Palouse Environmental Sustainability Coalition also took part, following megaloads along wintery roads in the dead of night to keep track of how often they stopped traffic and violated regulations. On one evening, monitors quickly spread the word when a megaload took out a power line, cutting electricity to 1,300 homes and businesses.

These groups also held information sessions, hosted film screenings about the tar sands extraction impacts, organized community members to attend the annual tar sands Healing Walks in Alberta, and did public outreach. Helen Yost, the core organizer of WIRT, inspired many to join the struggle by spreading information during the weekly Moscow Farmers Market and Climate Justice Forum radio show, through a WIRT Facebook page and website, and by canvassing and gathering anti-megaload petition signatures.

Protest and Blockade. Getting out in the streets was another core component of the struggle. To confront megaloads as they came through towns, WIRT instigated demonstrations and monitoring throughout Idaho and Washington and collaborated with FOC to host protests in Lewiston and Moscow, Idaho. In Moscow, where the majority of these demonstrations occurred, protesters met megaloads in the downtown area, night after night and during winter, holding signs and sometimes sitting down in the road and risking arrest. From 2011 to 2012, sustained protests met approximately seventy loads that traveled through Moscow on thirty occasions. There were thirteen arrests and citations during five protesting and monitoring events. The loads often came through town between 11:00 p.m. and 2:00 a.m.

Protesters and tactics in Moscow were diverse. Grandmothers composed a group of the protesters. The local Moscow Volunteer Peace Band played on occasion. One evening, women engaged in street theater. Dressed in formal gowns, they planned to enter and stall in a crosswalk, when the megaload came uphill, blocking its path. Despite the grandmothers' location outside the typical protest zone, the police seemed to know of their plan and arrived while the protesters were waiting for the megaload, preventing them from crossing the road. Moscow's mayor at the time, Nancy Chaney, was supportive of the resistance, writing letters to agencies and observing the protests, to ensure appropriate interaction between police and protesters. She gave a Mayor's 2012 Earth Day Award to the megaload protesters.

Despite the February 2013 ruling that the Forest Service has the jurisdiction to regulate megaloads on Highway 12, and a Nez Perce resolution of megaload opposition, the Idaho Transportation Department issued megaload permits to the transport company Omega Morgan. So, on August 5, 2013, approximately 150 Nez Perce met the first Omega Morgan megaload with a blockade on Highway 12 at their reservation boundary.

> Holding landing signals, women and children direct the megaload back from the Nez Perce reservation border. Approximately 150 other people whoop and holler, singing Nez Perce tribal chants and pounding drums. White, blue,

and red lights blare in the eyes of protesters, pointed from the line of police vehicles that escort the megaload. "Turn it back, Turn it around," people yell. Women, men, and children are on the frontlines. White non-Native faces are sprinkled through the crowd of Nez Perce. People are recording on their phones and cameras, making my fieldnotes on the evening possible three years later. A man holds a small girl, probably three years old, who shakes the blanket in her tiny hand in rhythm with the chants. Signs reflect a diversity of concerns —"grammas against megaloads," "climate killers," "blow your megaload somewhere else," "stop the machines of planetary rape— oppose the megaloads," "why work for murderers?" The drum and chanting are a constant presence. Nez Perce tribal police—approximately 10 cars, are spatially behind the protesters. (December 2016 fieldnotes on video taken August 5, 2013, https://www.youtube.com/watch?v=Gzu73gTEEeo)

Much of the organizing for the blockade took place on Facebook. Efforts of tribal members like Julian Matthews, who started the group Nimiipuu Protecting the Environment, helped convince the Tribal Executive Committee to take a stance on the megaloads. On that night, after much singing, drumming, and confrontation, eight members of the Nez Perce Tribal Executive Committee (NPTEC) were arrested and escorted away by tribal police, in what interviewees saw as a symbolic arrest.

Blockades continued for the next three nights, with twenty-eight Nez Perce arrested in total. In contrast to the first night, these arrests were what Paulette Smith described as "protest real," meaning violent. On these nights, the Idaho State Police forcibly arrested Nez Perce protesters in what Lin Laughy and Borg Hendrickson called "an ugly affair." Tribal member Paulette Smith was dragged under a megaload, and her daughter was punched in the face by an Idaho State Police officer. Reflecting two years later on the injustice of the event still brought emotion to Paulette's voice. She explained: "Our leaders got to have their hands in front of them [when handcuffed], they were walked off into this little cart thing, and, [. . .] from what I was told, they were let out on fifty-dollar bond, they really weren't arrested and stamped in. The second night, no, that's different. I have every tattoo on my body cataloged with Idaho State Police and Nez Perce County, I was fingerprinted; I was treated like a criminal, I was manhandled." Non-Natives attended the protests to stand in solidarity but were the first to be moved to the sidelines by police. Nonetheless, this support helped grow a foundation of collaboration between Nez Perce and non-Native environmental activists in the area, who continue to work together.

The Nez Perce blockade was a momentous occasion for the loose coalition of activists and organizations that had been fighting megaloads since 2010. It was not only a turning point in legal battles that eventually restricted the megaloads, but also a powerful example of solidarity and resistance that Nez Perce and non-Native interviewees remember with emotion. Like other instances of Indigenous direct action—Standing Rock, especially—state and corporate decisions to ignore tribal sovereignty and appropriate legal processes sparked the action.

The megaload story began to close one month later, on September 13, 2013, when Judge Winmill issued the megaload injunction on Highway 12. Over the next year, a dozen megaloads tried to take five alternate routes through Idaho, Montana, and Oregon. They met more grassroots and Indigenous protectors, protesters, and monitors, an Oregon lawsuit, state agency and corporate office occupations and meeting disruptions, and statements of opposition—all from dozens of groups, including the Coeur d'Alene, Shoshone-Bannock, Umatilla, and Warm Springs Tribes, and Indian Peoples Action in Montana. Approximately fifty direct encounters occurred, resulting in twenty-six arrests and citations. In summary, what oil companies hoped would be an easy and profitable plan became just the opposite. Idahoans and the Nez Perce would not allow their ancestral lands, wild places, and towns to become sacrifice zones or to support the sacrifice zone of Alberta tar sands exploitation. Their struggle moved beyond local activism as participants worked together across regions and in solidarity with people suffering the effects of extreme extraction and climate change everywhere.

Megaload Conclusion

Since 2013, Nimiipuu Protecting the Environment and various conservation and climate groups have continued to collaborate, building on the bonds and trust forged during the megaload fight. One of their central campaigns has been to remove four dams on the lower Snake River in order to improve habitat for salmon that play an important role in Nez Perce culture and nutrition. The Nez Perce Tribe also persists in its resistance to Big Oil, having issued a statement in support of the Standing Rock Sioux Tribe's opposition to the Dakota Access Pipeline. With assistance from University of Idaho professor Leontina Hormel, the Nez Perce led a study of the importance of the wild and scenic river corridor, as part of the Forest Service's court-ordered mediation with the Tribe. Despite this ongoing mediation, the Idaho Transportation Department adopted new megaload rules for Highway 12 in November 2016. Opponents sent hundreds of letters in protest. In January 2017, the Nez Perce Tribe, Idaho Rivers United, and legal group Advocates for the West reached an agreement with the US Forest Service that prohibited megaload shipments over certain dimensions on Highway 12 and resolved the 2013 federal court lawsuit. Highway 12, then, is protected from megaloads, but other roads remain at risk.

This case demonstrates how a diversity of motivations and tactics can be used together for powerful effect. Groups worked in collaboration, but also on their own, enabling spontaneity and persistence. They achieved their goals and brought national attention to their struggle (see Johnson (2013) and Zeller (2010) in the *New York Times*). It also reveals how a sense of injustice, informed by connection to place and feelings of solidarity with people on the frontlines of energy extraction and climate change, can facilitate working across lines.

Measure P in Santa Barbara, in contrast with the megaloads activism, is a case where working across lines was not as successful and where autonomous horizontal forms of organizing were eventually discarded for hierarchical political party organizing.

SANTA BARBARA: MEASURE P PROTECT OUR WATER

On January 28, 1969, a blowout occurred below Union Oil's platform in the Santa Barbara channel.[4] It was the worst oil spill the nation had seen and an ecological catastrophe. It was also a catalyst for progressive environmental regulations and the modern environmental movement. Thomas Storke, editor of the local *Santa Barbara News-Press*, wrote, "Never in my long lifetime have I ever seen such an aroused populace at the grassroots level. This oil pollution has done something I have never seen before in Santa Barbara—it has united citizens of all political persuasions in a truly nonpartisan cause" (quoted in Loomis 2015).

Forty-five years later, a small group of Santa Barbara residents came together to reinvigorate grassroots power to confront the oil industry. Meeting in late February 2014 as 350 Santa Barbara (350SB), a one-year-old chapter of the international climate movement organization 350.org, members listened as the two women who would come to lead the Yes on Measure P campaign, Katie Davis and Becca Claassen, proposed the ballot measure idea. The other six activists in the room agreed to support the effort and on April 5, volunteers around the county began collecting signatures to support an initiative that would ban the use of unconventional energy extraction techniques—fracking, acidization, and cyclic steam injection—in new wells. I was part of the campaign from the beginning and, in the next two years, conducted forty-three in-depth interviews with local activists.

In the next seven months, people who had never been active before mobilized around the issue; longtime political strategists and heads of Santa Barbara's many environmental nonprofit organizations weighed in; and Big Oil, perhaps surprised by the signature-gathering success, poured money into an opposition campaign. The confluence of these factors, combined with the demographic and political context of the county, informed the strengths and weaknesses of the campaign and the threats and opportunities that activists confronted.

Strengths and Opportunities

Activists behind the effort to qualify Measure P for the ballot successfully mobilized a broad base of not only supporters, but also people who would take time out of their day to stand on the corner, in front of the grocery store, and on campus to ask people to sign petitions. Throughout the campaign, over one thousand volunteers mobilized— more than for any other anti-fracking effort in California to date. More than this quantity, however, was the quality of volunteers' engagement. Accustomed to organizing through consensus-based decisions and without formal leadership, the originators of the measure were committed to horizontal leadership. They formed seven teams throughout the county. Each had one or two "team captains" who were responsible for training signature gatherers, collecting signature packets, and tallying signatures totals. In North County, residents alarmed by the oil production around them collected thousands of signatures. An undergraduate student and I were cocaptains for the area around the University of California, Santa Barbara. Each signature gatherer set personal goals and timelines. With this model, over twenty thousand handwritten signatures in support of placing the initiative on the ballot were collected in three weeks (see figure 9). Sixteen thousand—three thousand more than needed—were deemed valid because they were from registered voters. Meeting this threshold prompted the County Board of Supervisors to place the initiative, which became Measure P, before voters in November.

A sense of urgency contributed to the energy behind this early phase of the campaign. All core members had come together around the issue of climate change, many inspired by Bill McKibben's 2012 article in *Rolling Stone*, titled "Global Warming's Terrifying New Math," which argued that fossil fuels must be kept in the ground to avoid warming the

FIGURE 9. The Santa Barbara County Water Guardians pose next to 20,000 signatures submitted to the county elections office on May 5, 2014. Katie Davis, Janet Blevins, Becca Claassen, and I (all center) spent the morning frantically double-checking the validity of signatures before delivering the boxes to the county. Women's multigenerational commitment to the cause is evident. Becca is Janet's daughter and Becca's young daughter stands to my right. Photo by the author.

planet past two degrees Celsius. The failure of the 2013 United Nations climate change negotiations (COP19) contributed to our sense that *local* climate action was critical. COP19 made little progress toward a global, binding climate treaty. As a *Guardian* headline read: "COP19: The UN's Climate Talks Proved to be Just Another Cop Out" (Zammit-Lucia 2013). There was also an urgency to protect future generations. Becca Claassen and Katie Davis, two key organizers, felt so committed to the cause that they put their careers on hold to be full-time volunteer organizers. They both felt motivated by a desire to secure a livable future for their kids.

Another dimension of the urgency—why proponents pushed the initiative forward despite some more seasoned electoral campaigners' cautionary tales about low voter turnout in mid-term elections—was looming oil development. In 2012, 350 Santa Barbara had their first victory when the County Board of Supervisors required Santa Maria Energy, a local oil company, to purchase offsets for emissions above a threshold of 10,000 tons per year on its 136-well expansion. On the heels of this victory, activists became aware of an impending boom in unconventional oil production in the county, which sits atop the shale of the Monterey Formation. At the time, the formation was seen as a "black gold mine" of petroleum; federal energy authorities later downsized the estimated amount of recoverable oil by 96 percent (Sahagun 2014). In the short term, according to Santa Barbara County, applications for 903 wells were permitted, proposed, or anticipated, 89 percent of them using high-intensity techniques. In the longer term, Santa Maria Energy disclosed plans for 7,700 new wells (Nellis 2014). The emissions from these potential projects would eliminate 350 Santa Barbara's previous progress and cancel out emissions savings from all the lifestyle changes activists made to cut their individual carbon footprints.

Alongside their passion and sense of urgency, the core group of activists behind the measure had the time—what scholars refer to as "biographical availability" (McAdam 1986). They were retired, students, full-time activists, and people with flexible work schedules. They were at a stage in life where they were available for activism. With full-time activists Becca Claassen and Katie Davis, this availability was highly gendered, as both were, in fact, in a life stage when people are usually least available for activism. They were women in their thirties and forties with careers—Becca as a chiropractor and Katie as a tech executive—and dependent children. They *chose* to become available for activism. Thus, activists' passion to dedicate, or create, free time for the cause helped nourish a grassroots campaign that became a formidable force in local politics.

Threats and Weaknesses

Ultimately, the passion and energy of activists and high voter approval ratings (57 percent) in summer of 2014 proved insufficient to pass Measure P on Election Day. Based on interviews and my own participation in the campaign, I argue that entrenched oil industry power, insufficient groundwork before the campaign, and the change in tactics brought on by collaboration with the Democratic Party shaped the outcome.

The oil industry has a long presence in California and Santa Barbara County, where it pioneered offshore drilling in the late 1800s. In 2014, it continued its tradition of using money to shape politics (see Molotch 1970). A group called Californians for Energy Independence, whose donors included Chevron, Aera Energy, and local oil companies, surprised proponents by funneling over six million dollars to opposition efforts. Measure P proponents raised over four hundred thousand dollars—outspent more than fifteen to one. This unequal fundraising, combined with the fact that many contributions to proponents arrived late in the campaign, inhibited outreach efforts.

One effect of low funding for outreach was that some voters assumed that the opponents represented their interests. As Hazel Davalos, organizing director for Central Coast Alliance United for a Sustainable Economy (CAUSE), which supported the measure, explained, Latinx voters in North County frequently heard—on local and large-scale media platforms, including Univision and Telemundo—"No on P" messaging in Spanish from North County Latinx spokespeople. Latinx individuals comprised 43 percent of the county population of 424,896 in 2010, and 70 percent of the 99,553-person city of Santa Maria, North County's largest city. Hazel recounted that people said, "Oh they are actually advertising to us. They must be the campaign that has our interests at heart." Some of these ads focused on how loss of tax revenues from oil would hurt local schools that predominantly working-class Latinx communities—composed largely of agricultural workers—depend on.[5] Yes on P not only had fewer resources for media outreach but also made a costly mistake when nonresident campaign staff bought ads on Spanish-speaking radio in South, but not North County. Not having as large of a base in North County, Yes on P spokespeople were typically based in South County. Therefore, they were not people that most North County residents recognized.

The sheer quantity of No on P messages, which centered on loss of jobs and tax revenues, was also insurmountable. Proponents spent their energy countering industry lies, rather than educating the community about the water and health risks of fracking—a message that is as relevant to Latinx agricultural workers as it is to white tourist sector employees, who both depend on the environment for their livelihoods. Creating doubt is a powerful tactic in and of itself, as evidenced by the fossil fuels industry's efforts to manufacture doubt about climate change (Oreskes and Conway 2010). Central points around which doubt emerged were whether the initiative applied to existing oil wells (it did not) and, related to this, if the county would be at risk of lawsuits. A county staff report (County of Santa Barbara Staff 2014) with incorrect information exacerbated this situation and was cited by opponents throughout the campaign despite the report's rejection by the County Supervisors. In Becca Claassen's view, the county suffers from regulatory capture by the oil industry—staff in Santa Barbara County's Energy Division depend on the oil status quo for their jobs. For this reason, they are inclined to keep oil going and are therefore "captured" by the industry. Finally, industry arguments overstated the jobs and taxes supplied by the industry. Oil and gas production accounted for only 1.07 percent of county jobs and 2.65 percent of county property taxes (County of Santa Barbara Staff 2014) and directly threatens the county's largest economic sectors—agriculture and tourism.[6] The measure exempted existing production, and therefore, would have had little effect on existing jobs and tax revenues.

Insufficient groundwork leading up to the campaign was a primary weakness of its proponents' efforts. They were a small grassroots group, that, on hearing that their efforts would be negated by oil expansion, decided to take electoral action in a much shorter period than is customary. More established members of Santa Barbara County's environmental community cautioned the proponents about the difficulty of passing a fracking ban in a mid-term election, which are known to have low voter, and especially low progressive voter, turnout. In interviews following the election, these established environmental figures and party leaders also explained that the rapidity of the campaign inhibited relationship building with local leaders, labor unions, key spokespeople, and potential donors. This lack of connections likely contributed to surprising endorsements of "No on P" by the generally progressive local newspaper, the Santa Barbara Independent—though a coalition of editors wrote against this endorsement—and by public safety groups such as the Santa Barbara Fire Fighters Local and Police Officers Association. As described previously, the short timeline made it difficult to build relationships and cultivate support in the Latinx community. More preparation time, explained Linda Krop, chief counsel of the Environmental Defense Center in Santa Barbara, might also have helped make the initiative clearer for county staff. Greater clarity may have prevented the dissemination of false information and confusion:

> It goes back to the starting early. When I've written initiatives before, I will often run them by county staff and counsel, and say here's what we're trying to do. [. . . Agency staff can't take a formal position, but] they can look at it and say, "What do you mean by this?" or "We are concerned about how this might affect well maintenance," and then we can have that conversation and resolve it before it hits the streets, and then we don't have the county's opposition. So, all of those kind of preliminary things didn't happen.

Many of these established groups, though skeptical and cautionary about the proponents' plans, came on board as momentum built. On May 12, when the initiative qualified for the ballot, these groups were pleased, and a bit surprised by the overwhelming success of the signature efforts.

The local Democratic Party decided to endorse the measure soon after it qualified for the ballot and offered to partner with the campaign. The measure's proponents, overwhelmed by the input from people seen as experts in local elections and environmental politics, and tempted by the idea that securing the support of loyal Democrats and Independents who vote Democratic would be their best chance to win, agreed. Though Santa Barbara County leans Democratic, in October 2014, only 40 percent of voters were registered Democrats. Aligning with the Democratic Party may have dampened bipartisan support for the measure, whose focus on water quality had potential as a bipartisan message.

Interviewees also argued that turning the campaign over to the Democratic Party had the effect of changing the structure of organizing, something I experienced as a participant in the campaign. As Arlo explained,

> [In fall 2014] I found that the coalition of folks [. . .] leading this campaign had changed drastically [from the spring], and it suddenly included largely folks from the Democratic Party of Santa Barbara, and [. . .] it was just a much different beast. [. . .] We ended up putting so many of our volunteer hours—us being all the folks who helped get Measure P on the ballot—into talking about issues that weren't Measure P or talking about [Democratic] candidates. And it felt really frustrating because yes, the Democratic Party was allowing use of their office downtown, but otherwise, like what were we really getting out of them? [. . .] There were times when it felt like they were getting a lot more out of us than we were getting out of them.

Whereas proponents had cultivated distributive leadership through the team captain structure in the signature phase, during the lead-up to the November election, this original structure was disbanded and replaced with phone banking and precinct walking in which volunteers were instructed, through a hierarchical leadership structure, to follow scripts explicitly connected to the Democratic Party to communicate with voters. While volunteers had previously focused on their own neighborhoods, in this model, they were distributed throughout the county, regardless of their residence. This new approach, combined with burnout from the signature effort, disillusioned some key volunteers, myself included. All volunteers, no matter what role they had played in the signature effort, were expected to return all data to the Democratic Party, which took primary responsibility for compiling data, developing timelines, and determining which voters volunteers contacted. Becca Claassen co-originated the campaign, and though she had the title of campaign committee chair, she felt disempowered by the new campaign structure. If she were to do it again, she explained, she would have advocated

> starting voter contact much earlier, less phone calls and more door-to-door, more face-to-face, more empowerment and actual relationship building, not considering everybody on your list just a volunteer. Sure, there are those people who just want to be told what to do [. . .] but there are other people who have valid concerns and opinions and know their neighborhoods better than you do. We were encouraged [by the Democratic Party] to just like tell the volunteers what to do, make them stick to the script. It felt very hierarchical and top-down, not empowering, if anything it was disempowering.

Signature gathering team captain Juan agreed, saying in 2016, "Maybe we have PTSDP"—posttraumatic stress disorder from Measure P.

The collaboration did have positive elements. It allowed proponents to use Democratic Party office space, to have access to precinct and phone banking databases, and to have their measure on party literature. Together, Democratic Party and "Yes on P" volunteers made hundreds of thousands of phone calls and knocked on thousands of doors in an unprecedented field campaign in the county. Yet alongside these benefits came a change in the spirit of the campaign that many core organizers regretted. They felt that the Democratic Party benefitted more from the popular movement proponents galvanized than the measure benefitted from the Democratic Party. They looked to the success of a 2014 fracking ban in San Benito County, where the grassroots originators of the campaign maintained autonomy from political parties, as an example of a different method for carrying out the campaign. In 2016, Monterey County also successfully passed a fracking ban with an autonomous grassroots campaign. Monterey County has a higher percentage of registered Democrats than Santa Barbara County but also produces more oil and faced similar opposition funding.

The vote reflected sharp demographic and regional divides in the county. Overall, 39 percent of voters were in support, comprised largely of students and residents of the city of Santa Barbara. Nearly 80 percent of voters in North County, largely Latinx and conservative-leaning white voters, were against the measure, compared to 40 percent in South County. These results were also affected by epically low voter turnout.

Low voter turnout is common during mid-term elections because these elections lack the surge in interest and information that presidential elections spur (Campbell 1987). In Santa Barbara County, only 58 percent of registered voters cast votes on Measure P. Nationwide, it was the lowest voter turnout in seventy-two years; 36.4 percent of eligible voters (Alter 2014) and only 19.9 percent of eighteen- to twenty-nine-year-olds voted (CIRCLE 2015). Some reasons for this low overall voter and low young voter turnout include more restrictive voting laws, a lack of competitive races, and a lack of outreach to young voters (see CIRCLE 2015). In Santa Barbara, voters are not encouraged to register to vote on election day, and the 2014 gubernatorial election, where incumbent Jerry Brown won 60 percent of the vote, was not competitive. These factors may help explain why UCSB students and Isla Vista, California, residents, the strongest supporters of Measure P, with 80 percent yes votes, had only 15 percent voter turnout.

Post–Measure P Outcomes

Though Measure P failed, it represents an impressive example of grassroots mobilization, particularly in the signature-gathering phase. With more preparation time, less opposition from the oil industry, the *Santa Barbara Independent*, and county staff, and greater voter turnout, the results may have been different. While heads of established environmental organizations in Santa Barbara feared the defeat would be a setback for local environmental politics, these fears did not materialize.

In spring 2015, the Santa Barbara County Board of Supervisors approved the state's most stringent carbon emissions levels for oil projects and funded a feasibility study for community choice aggregation, which could increase renewable energy in the county. In November 2016, the board rejected an oil project that had much stricter controls than the Santa Maria Energy Project, described previously, which the same board *approved* in 2013. In the 2016 election, former environmental attorney Joan Hartmann won the contentious Third District Supervisor seat, which straddles North and South County, maintaining a three-to-two progressive majority on the Board of Supervisors. Her opponent, Bruce Porter, received over $60,000 in funding from energy producers. In 2021, Katie Davis informed me that not a single major oil project has been approved in Santa Barbara County since the Measure P campaign.

Positive energy outcomes occurred in other locations as well. In the 2014 and 2016 elections, there were multiple ballot and legislative efforts to ban fracking throughout the United States. New York State banned fracking in 2014. With the addition of Monterey County, California had six counties that had banned fracking by the end of 2016. Monterey County's "Measure Z" is more restrictive than Measure P, banning intensive techniques, wastewater injection, and *all* future conventional wells. In 2021, California governor Gavin Newsom moved to ban new fracking permits beginning in 2024.

In summary, Measure P began as a grassroots effort that relied on relational and horizontal organizing. These tactics changed with leadership by the Democratic Party. Party organizing foreclosed possibilities for creativity and spontaneity, which, even at the beginning of the effort, when it was still controlled by the grassroots, were tactics that received little support from established environmental organizations. Established organizations faulted the campaign for not having built prior relationships with established environmental groups, people in power, and representatives of labor and public safety groups. In contrast, most grassroots activists identified their main weakness as not having built relationships with the Latinx community. Max Golding said that if he could go back in time to when he cofounded 350 Santa Barbara, which started the Measure P campaign, "first [. . .] we would've said, 'What do we want this movement to look like?' It would've been a more deliberate question and [. . .] we all would've said, 'We want Latinos, we want this to actually represent who the fuck lives here, and we would have talked to [Latino organizations].'" The power of the oil industry to spread doubt and draw on its long presence in the political economy of the area also inhibited identification and communication of common values between South County activists concerned about climate change and North County residents and Latinx people fearful of losing jobs and tax revenues. Had they maintained their grassroots organizing structure, Measure P proponents may have been more successful communicating their strongest message—the risks of fracking—to key constituencies, even in the short time frame of the campaign.

The next chapter draws lessons from these struggles, highlighting similarities, differences, and needed work for building more inclusive movements.

CHAPTER 8

Lessons from Measure P and the Megaloads

Native–Non-Native and Latinx-White Coalition Outcomes

This chapter examines the differences and similarities between the Measure P and megaload campaigns, concluding with reflections on how to strengthen movements' commitment to diversity and inclusivity of participants, tactics, strategies, and motivations. It illustrates how political economy and sense of place shape how activists can frame extreme energy extraction. Bringing people together to resist oil infrastructure that is conceptualized as weapons of mass destruction is much easier when local people's jobs and tax bases are not grounded in that infrastructure. Beyond framing, I also find that the rigidity of the movement is important for influencing working across lines. Coalition work was less successful when activists tried to unite around one message, set of tactics, and way of building relationships and more successful when groups retained the freedom to employ different messages, tactics, and ways of movement building, in line with each group's priorities. Both of these factors confirm findings from other chapters of this book, including the importance youth place on intersectionality—recognizing and valuing difference, whether in identity or in campaign strategy—and the importance Idaho activists place on framing the natural gas company as lacking integrity and accountability, in other words, as immoral.

FIGURE 10. A deserted Highway 12, November 2015. Photo by the author.

KEY DIFFERENCES

The megaload and Measure P campaigns took place in vastly different settings—one along rural Idaho highways (see figure 10) and in Idaho political circles with very little experience interacting with oil corporations, the other in diverse agricultural and tourist cities in a county that has been producing oil for over a century. Living in Idaho, it is easy to forget about fossil fuels beyond trips to the gas station. Living in Santa Barbara County, oil has a constant presence. Derricks dot the ocean and agricultural horizons that many people see on a weekly basis. When I lived in Santa Barbara County, a clear view of the offshore oil rig, Platform Holly, troubled me on a daily basis—every time I walked to my front door (see figure 11). The presence and power of oil in these places is important in countering the weight of stereotypically friendly and unfriendly environmental politics in both states.

While Santa Barbara County, and California more generally, have progressive environmental protections, the entrenched power of oil and the resulting regulatory capture make it a difficult place to challenge oil corporations.[1] During Measure P, there was a sense among a sizable portion of the population that oil production is heavily regulated in the

FIGURE 11. Platform Holly disrupting a sunset. Photo by the author.

county, which makes it relatively safe, that it provides high-wage jobs, and that it supports local taxes. "No on Measure P" arguments that inflated the numbers of jobs and taxes that the industry provides, and incorrectly stated that the measure would shut down all onshore oil production, sought to create fear that the county's beneficial relationship with oil would come to an end. The intimate relationship that the county has with oil made it more difficult for activists to portray the oil companies as inherently evil, a feeling that, in contrast, was common among Idaho interviewees. The "Yes on Measure P" campaign, for example, specifically exempted existing oil operations from the measure. Megaloads opponents were less willing to compromise.

As protest signs like "Megaloads of Death" demonstrate, megaload protesters had clear ideas about the moral implications of megaloads, ideas that they stated publicly. In their view, these large transports were fundamentally wrong and not benefitting anyone, certainly not Idahoans. Thus, unlike in Santa Barbara, where activists tried to appease the local energy economy by exempting existing operations, Idaho protesters publicly opposed all elements of the megaloads. Their lack of dependence on oil corporations helped them overcome a political environment that is friendly to industry and hostile to environment protections (see chapter 2).[2] The clarity of their message also attracted widespread support; as Borg Hendrickson explained, "One motivator for doing something like this is the growing sense of indignity, the lack of justice, [. . .] and I have to make this right. I just must make this right; this is so wrong. You just build that sense."

The physical qualities of the fossil fuel infrastructure in both states contributed to these divergent perceptions of industry as evil versus industry as a member of the community. Megaloads were truly humongous, brilliantly lit pieces of machinery that traveled through downtowns and on roads in front of people's homes. Their presence was new and stunning, inspiring people to take a stand. In Santa Barbara, in contrast, the daily visibility of oil infrastructures likely desensitized residents to the threat. The oil rigs are also not in your face—they are visible in the distance and away from homes, shining like diamonds at night, nothing like a megaload next to your favorite coffee shop.

The political, social, and physical contexts I describe also contributed to Idahoans' feelings that industry was encroaching on their way of life and made it easy to imagine disastrous accidents, drawing from accounts they had heard of in other locations.[3] Tina Fisher, voicing her fears in relation to natural gas development in Idaho, expressed this widely held value about the quality of life: "On our day off, if we want to sit on the back porch and watch the world go by, we should be able to do that instead of having to wonder, well, what's in the air that I'm breathing? [. . .] I cannot imagine how helpless those people feel in Pennsylvania and other parts of the country [with fracking] where their kids wake up in the middle of the night with bloody noses. And the kids can't play outside." In Santa Barbara, people had coexisted with oil for decades, and though they were familiar with disasters in the form of two oil spills, these never directly harmed human health. Their imaginations of disaster, then, based on actual events, are likely devastating for sea life, but not for the majority of county residents.

The meaning of *quality of life* was heavily informed by interviewees' connection to place, something that defines Native–non-Native coalitions in diverse settings (Grossman 2017; Willow 2019). This connection was a powerful force in individual and organizational collaborations. When I asked Meryl Kastin why she cared about the megaload issue, the first thing she said was, "You've driven the road [Hwy 12]?" I had and recalled the more-than-hundred-mile stretch of winding highway, looked over by tall snow-dusted pines—the only other life forms I encountered during my last trip on the road. I could relate to Meryl's implication—Highway 12 is a special, peaceful part of the world. Lucinda Simpson, a Nez Perce elder, also gave a place-based explanation of why the issue was important to her, illustrating the Nez Perce's particular connection to the more-than-human world: "The nontribal don't see the importance of these things [strong voice, emotion laden], but the eel is a part of the food chain. If we didn't have that eel there, the fish wouldn't get fed and we wouldn't have our natural omega vitamins and fish oil and so forth, [. . .] and there used to be about twenty-nine different roots that we had here on this reservation and we are down to about seven."

These excerpts also illustrate a shared sense that the places disrupted by the megaloads had to be protected as commons, as wild and ancestral lands that belong to the public. The wilderness of Idaho (which comprises 14 percent of the state) belongs to the public, and the Nez Perce reservation belongs to the Nez Perce. Jeannie McHale alluded to this feeling: "I just love Idaho. That was another emotional aspect of the megaloads, was that they were just, [heavy emotion] they were just such an affront to our beautiful state. [. . .] It deserves to be protected by everybody because we have a lot of land that belongs to everybody." Though Santa Barbara activists could employ a message about the atmosphere and water as commons, they could not draw on the notion of the land as commons to the same extent as Idaho activists. The majority of onshore oil extraction in Santa Barbara County takes place on private lands and out of sight of most residents.

One way in which Idaho communities found inspiration through place was through NIMBYism. As I explain in chapter 4, interviewees in southwest Idaho were often motivated by a threat to their home and then expanded their concern to larger areas. This resonated with Helen Yost, who reflected, "People are NIMBY, you've got to admit they are, but if that's what it takes. You know there is nothing wrong with defending your place. [. . .] I mean that's the only thing you can really feel anything for anyways, your own place, be it a region, which is how we tend to look [. . .] or a road."[4] To borrow from Geertz, "No one lives in the world in general" (1996:262)—people live in and feel connection with certain places.

Interviewees throughout Idaho had moved there or chosen to live in the state for reasons that emphasized the natural environment. "We moved to Idaho thinking that clean air, clean water, you know," said Alma Hasse. "We love the area, we are of the area. [. . .] [Our kids and grandkids] love the place, the rivers," explained Lin Laughy and Borg Hendrickson. When Sherry Gordon moved from Humboldt County, California, to Idaho, she told her friends, "It's gorgeous there [in Idaho]." Rod Barklay, whom I interviewed in Sandpoint, Idaho, took his first backpacking trip when he was eight or nine years old in the Selway-Bitterroot Wilderness of Idaho. In his sixties at the time of our interview, his love of wilderness had grown throughout his life, informing his support of activist groups like Wild Idaho Rising Tide.

Like Rod, many interviewees in central and northern Idaho drew on notions of protecting places in their account of their participation in the megaload campaign. Their connection to place had a particular quality. As Lin Laughy, who had lived in the Lewiston, Idaho, area since 1948 explained, "It's a sense of place. You are involved if you are emotionally and spiritually involved with your place, and [when] somebody wants to come in and trample it, then you get upset." Helen Yost's whole journey into climate justice activism and her connection with Idaho grew from her previous connection to Alaska, where she lived in the 1980s. The Exxon Valdez spill in Alaska in 1989 sparked her activism: "[The Exxon Valdez spill] broke my heart, and it was the first time I have ever grieved. Like here, I found this remote beautiful wild place, lived in this little, tiny fishing town [. . .] to find a wild place like that and to know it over the course of, I don't know, five to eight years, and then to just have it be totally trashed by the oil company, it just tore my heart out." Her grief at the destruction of a place she loved spurred her to dedicate her life to activism when Exxon once again "pointed their headlights" at her special place, which, in the 2000s, was northern Idaho. "You are not going to ruin a second wild place on me. Like I could not bear it, and you guys [Exxon] still owe me for the last time around. [. . .] So I just jumped right in. I guess it was sort of a personal vendetta," said Helen.

Nez Perce interviewees had deep connections to place—their ancestors had lived on the land the megaloads threatened. The Heart of the Monster, where the Nez Perce people come from, is located just a few hundred yards from Highway 12. The megaloads also, by supporting extreme energy extraction, exacerbated the threats that climate change poses for the Nez Perce. Paulette Smith explained that climate change was "affecting how we are going to survive, or how we're going to exist, simply." Lucinda Simpson continued Paulette's point:

> We haven't had our normal seasons for the Indian people. Our foundation is built on water; the Nez Perce tribal people used to drink water first thing. First thing they would get up in the morning they would have a cup of water. [. . .] And that water is not clean any longer. [. . .] We, of course, our snow, our food chains, our gathering chains have all changed. The heat has damaged many of our roots and our berries and of course our fish. And the dryness and the heat has caused these fires in our areas, and we are probably

going to be seeing a lot of dead trees and more fires, a lot of trees' limbs will be falling off these trees when it starts to freeze. This winter we are going to see a lot of damage.

Lucinda saw the megaloads as just one more threat to the Nez Perce way of life, a way of life tied to the land, plants, and animals that shape Nez Perce culture. These connections paralleled the sentiments of Michael Cordero, an elder in the Chumash community in Santa Barbara County, California, who had grown up in Santa Barbara. His resistance to extreme energy extraction was informed by the fact that energy companies had disturbed places that the Chumash consider sacred.

Non-Native Idaho-based interviewees also drew on long-term attachment to place in their resistance to extreme energy and other forms of resource extraction. Shelley, a non-Native activist in southwest Idaho, highlighted the importance of history in a place when she explained that her daughter's great-great-grandparents had homesteaded in Eagle, Idaho: "A lot of people lived there [in Eagle] for generations. [. . .] That this [natural gas development] now could wreck everything that 150 years of history, you know, it's just, it's just so unjust." Cass Davis, who was a fifth-generation timber worker (working in a sawmill for a period after high school) and had been involved in many environmental and social justice campaigns before the megaloads explained,

> I knew that logging had fucked up my hunting and fishing back in the [Silver] Valley and places I knew, that it destroys cutthroat habitat. I knew all this stuff and it wasn't because I went to college; it was because I fish and I hunted and you can see. And as many people as I hung around who were loggers [. . .] all of them would say, "Oh, it [logging] doesn't hurt anything, we've been doing this for years," and I am like, "How in the fuck can you figure that?" I mean I grew up just like them, been raised the same way. [. . .] I seen all the trees disappear. I seen the crick [creek] used to run here, and now dries up completely during the summertime. "How in the fuck can you say that this doesn't hurt anything? What the hell is wrong with you guys?" And so, it wasn't hard to get me involved [in activism]. [. . .] The timber industry hated me for it because I wasn't a guy from back East. I wasn't a hippie; I wasn't their stereotype. I was salt-of-the earth Idaho, and I was saying no, this is stupid; we don't need to cut these.

Thus, for white Idahoans, Nez Perce individuals, and a few California-based interviewees, attachment to place played a key role in their activism. Throughout Idaho, it was common for interviewees to ask me if I had been to the places they spoke about. In parallel to the toxic tour that I took with Alma in southern Idaho, where she showed me natural gas wells and risks to the community, upon her suggestion, Helen and I drove along the megaload routes.[5] As we drove, she pointed out where she had been arrested and where the trees had been trimmed to make room for the loads. There were few suggestions among Santa Barbara interviewees to visit particular places. Indeed, Santa Barbara interviewee Katie Davis explained how her intense dedication to activism came from concern about climate change, not the environment: "I want clean water and clean air. But I wouldn't like devote my life to it. [. . .] If climate change wasn't an issue, that [environmental issues] wouldn't be where I'd spend all my time." The climate crisis, rather than a sense of attachment to a particular place, was Katie's main motivation. This characterizes the motivations of many youth interviewees in Santa Barbara as well. As students, they were temporary residents; they rarely expressed connection to Santa Barbara as a motivator.[6]

Finally, the strategies employed in these campaigns played a large role in shaping activists' experiences and capacities to engage in the type of organizing they desired. As a ballot measure, Measure P had a shorter timeline than many legal suits, and, since proponents started so much later than is customary, the timeline was even shorter. From start to finish, Measure P was about an eleven-month campaign. The megaload campaign went on for seven years, and while it is unlikely Highway 12 will ever be at risk again as a megaload route, other highways still lack protection. Megaload proponents had a lot more time to work on their relationships than did the proponents of Measure P. There was more time for a large and not necessarily coordinated network to develop organically as more and more people became concerned about the issue and used different methods to challenge the loads. People were literally "popping up," as Helen would say, everywhere and across state lines. In contrast, proponents of Measure P had one centralized group and message on which oil companies could focus their opposition, like a laser beam.

KEY SIMILARITIES

Despite the differences in ties to the oil industry, immediacy of fossil fuel infrastructure, sense of and attachment to place, and tactical timelines, the megaload and Measure P campaigns had significant commonalities. These struggles were rooted in the nature of resistance to extreme energy extraction. This book argues that people resist fossil fuels primarily for the risks they pose to basic needs that most people value—healthy land, air, water, climate, and community. Not every campaign can appeal to all these shared values, but each can appeal to at least one. Organizers in these campaigns and across my research sites noted how these issues had the capacity to draw new people to the movement and cross demographics. The following quote from Hazel Davalos was typical: "The grassroots energy behind the [Measure P] campaign, it really brought out a lot of relatively new faces; it wasn't like the usual suspects, [. . .] It really started to galvanize an environmental kind of network that I didn't really see here [in Santa Maria, CA] before that."

Perceived threats created a sense of emergency among interviewees in both states, leading them to draw parallels with World War II. Sharon Cousins, while understanding why some people could not protest at 2 a.m. against the megaloads because of having to wake up for work, nonetheless felt that the megaloads, when fully understood, were an emergency that demanded action from everyone. As Sharon said, "I can only think [people who didn't protest] just don't really understand what [the megaloads] are—it's like if you lived in some town in Europe and had trucks loaded with supplies to build the gas chambers at Auschwitz going through your town. [. . .] If you really see what is going on, you can't not take a stand." Katie Davis had a similar perspective about climate change that inspired her to start organizing in her community of Goleta, California. She had just seen Al Gore's *Inconvenient Truth* with some German friends. These friends had grandparents in Germany during World War II, and a rift had developed in their family about why the grandparents had not done more to challenge Nazism. Katie explained that, after watching the film, her friend "wondered out loud if climate change was going to be something like that for us, something that our kids are going to ask us about: What did we do? Why we didn't do more when we still could, you know? I just remember that really hitting home because I had two young kids at the time."

Illustrating just how corrupt many megaload interviewees felt their government was—in parallel to CAIA's emphasis on lack of integrity and accountability that I describe in chapter 3—Jeannie McHale criticized the Port of Lewiston, which allowed the megaloads to be shipped to Idaho. "They could be hauling dead babies from Nazi concentration camps," said Jeannie. "The Port of Lewiston would welcome them in and use Idaho taxpayer dollars to fill the potholes in the road after they've passed through." Jeannie emphasized how the state of Idaho was not only willing to let horrendous practices happen but was also happy to use public resources to subsidize these practices—to subsidize corporate profits at the expense of people and the planet. Sense of emergency for people and/or the planet is like a river all activists are dunked into at some point in their journey.

The shared values underlying both campaigns inform another important similarity. These values did not simply disappear when Measure P failed and when the federal ruling barred the megaloads. Having brought people together who had not worked collectively before, these values laid the groundwork for future collaboration. Thus, both efforts enabled movement building that made communities stronger and ready to organize together on other issues. The megaload campaign strengthened the coalition that, upon the conclusion of the effort, worked to remove the lower four Snake River dams for the well-being of salmon and other fish, ecosystems, and the Nez Perce. "It brought a lot of people together that hadn't known each other before and then found out that they shared this interest in protecting the planet. [. . .] So, it built long-lasting friendships and [. . .] continued effort," said Jeannie McHale. "In a really screwy way," explained Brett Haverstick, "the greed of and destruction of the Alberta tar sands and the megaloads strengthened local communities."

In Santa Barbara County, Rebecca August used the skills and networks she developed during Measure P to form a new group, Safe Energy Now. They began organizing in the city of Buellton, where little activism on energy issues existed before Measure P. Janet Blevins's efforts to raise awareness about Measure P in the city of Lompoc led her to build many friendships and become active in groups like the NAACP and League of Women Voters, with whom she continued to organize long after 2014. The momentum of these campaigns then, kept flowing in different forms as activists continued to rely on the relationships that they built or strengthened while working together.

Despite their different outcomes, both campaigns benefitted from participants' shared sense of emergency, and both facilitated long-term relationship building. Future collaborations grew out of these relationships that were rooted in the shared values of protecting communities that originally motivated megaload resistance and Measure P. With the differences and similarities of each campaign in mind, the following section considers interviewees' best practices for working across lines.

REFLECTIONS ON WORKING ACROSS LINES: BEST PRACTICES

Engaging in these struggles equipped interviewees with ideas about how to be more effective in future efforts. Their reflections contained

insights for working across lines. Some developed their ideas throughout decades of organizing and in consultation with political philosophy.
Others had just begun to develop their theories, drawing on what went
well and what did not go well in their latest organizing experience. Prioritizing diversity and inclusivity of participants, tactics, strategies, and
motivations was common and important for success in both cases. To
achieve diversity and inclusivity, interviewees advocated bringing stakeholders together—"convening a table"—and working in a horizontal
fashion. The legal success of the megaloads and the electoral failure of
Measure P provide evidence for the utility of these methods. In this section, organized by theme, I draw on evidence and perspectives grounded
in both campaigns to discuss methods for effective organizing.

Convene a Table and Keep It Horizontal

Tables are a common metaphor in organizing. The question of who is at
the table is of prime concern. As detailed in chapter 6, maintaining a seat
at the table of power is a goal of some groups. Daraka Larimore-Hall,
chair of the Santa Barbara Democratic Party, explained that convening a
coalition through the "model of tables" was one thing he would have done
differently in the Measure P campaign: "I'm a big fan of the sort of model
of tables, of convening coalitions. [. . .] You've got to have stakeholders at
the table, you've got to get some buy-in. So, what I would have preferred
was [. . .] let's get the EDC [Environmental Defense Center] and the Sierra
Club and other environmental groups together with some labor folks [. . .]
and some of our electeds [elected officials], the Democratic Party. Let's sit
around the table with the folks that were driving the issue [. . .] and talk
about what we'd like to achieve." Interviewees in both places agreed that
convening tables with stakeholders was important. There was, however,
disagreement about which stakeholders most deserved to be at the table,
especially in terms of the power that they had, and how they conducted
themselves once there.[7] When 350 Santa Barbara was a new group, it had
to work to understand what tables existed in Santa Barbara and how to
have a seat. Becca Claassen and Colin Loustalot, two cofounders of the
group, "crashed" a local environmental coalition meeting as their first
meeting, in an effort to find a table. Across my research sites, new grassroots and radical groups had had experiences of not receiving invitations
to tables, or, once there, did not feel like their voices were heard.

In Santa Barbara, tensions between grasstops or established
organizations—many of the groups that Daraka mentioned—and the

grassroots proponents of Measure P, known as the Santa Barbara County Water Guardians (composed of 350 Santa Barbara plus a broader base of individuals [hereafter Water Guardians]), were grounded in the Water Guardians not having invited the grasstops to the table early enough. In Dave Davis's words, Measure P "came blowing out of the ground" without convening a table beforehand. While many of the grasstops groups were slow to endorse, they did so when the initiative qualified for the ballot. Because the Water Guardians did not decide to attempt the ballot measure until February 2014 (and had until May 2014 to gather 13,200 signatures of registered voters), there was little time to convene a table. If the Water Guardians had convened a table with the grasstops organizations ahead of time, they would have been advised by grasstops staff, as they were advised at later stages, not to attempt the initiative. According to representatives of the grasstops groups, it was too late, the wrong year, and too hard. Most did not think the signature effort would succeed. Thus, there may have been an impasse if a "table" had been convened before 2014. As I highlight in the preceding section, the Water Guardians felt a tremendous urgency to go forward with the effort, so it is unlikely they would have changed their minds.

In terms of outcomes, inviting grasstops to the table earlier in the campaign may have helped Measure P secure funding earlier, which could have allowed more effective spending. However, in grassroots activists' view, the real problem was not having convened a table with representatives of conservative and Latinx communities in North County, as well as other groups like public safety and labor unions. Note that Daraka did not mention conservative, Latinx, social justice, or student groups in his advice about convening a table. His omission of these sectors of the community does not mean that Daraka and representatives of the environmental community did not recognize the importance of building relationships with these groups—other parts of their interviews illustrate that they clearly did. What his omission does illustrate is which groups do and do not immediately come to mind when grasstops organizations in Santa Barbara think about beginning a campaign. In other words, it illustrates the priority groups to invite to the table. As Daraka went on to say about the table strategy, "It's not perfect and [. . .] can end up being kind of elitist [. . .] but I think it is a necessary component of progressive politics." Indeed, my interviews and fieldwork demonstrate that having a relatively powerless seat at an elite table characterized grassroots activists' experiences during Measure P, which eventually convened a table of elite environmental organizations and the Democratic Party, which took control of the campaign.

A good illustration of the character and atmosphere of the table that was eventually convened was the post–Measure P meeting. Following the election, proponents gathered around a huge wooden table in a community space in Santa Barbara. The following account is based on my notes from the evening, November 18, 2014.

> The grassroots originators of the campaign were the minority in the room of forty-one people—twenty-four women and seventeen men, mostly over forty years old. Only two people of color (women) were there. The Measure P campaign manager presented his assessment of the campaign, and then individuals gave their feedback. The majority of feedback was negative and fearful of the failure's ramifications. One woman said, "I'm not surprised. Why were we surprised? It was a bad idea to do this campaign because you lose a lot. It will be harder now." In a gendered comment about how passion is not an effective component of organizing, a man said, "Moving forward on passion without ability to deliver a win is counterproductive and actively sets us back." Katie Davis's views represented grassroots activists' sentiments, as they felt the effort was a real success in movement building and a necessary step toward addressing climate change. She explained, "The story is that two of three [California fracking] bans won; we have a lot of volunteers. There is a lot of good that came out of it even though we lost. If we wait for it all to be perfect, this planet is not going to be habitable. This requires new things and not being afraid and mad that people are willing to try new things." As most attendees continued to advocate for future narrow focus on keeping local Democratic elected officials in power, and using the "commodity" of the Water Guardian's email list to this end, Janet Blevins, also a grassroots activist, interjected. Showing the sharp contrast in grassroots' views, she said, "We'll be dead before we get enough [progressive people] elected to make progress on this. We don't have time to build a political elected alliance."

The tensions expressed during this meeting and the privileged social locations of the people around the table demonstrate that convening a table is not enough. Valuing diversity, inclusivity, and accountability to and leadership by those most affected by a problem—core elements of climate justice—were absent. Those most affected by the fossil fuel extraction occurring in Santa Barbara County are people who live close to onshore oil production in North County; the poor and marginalized in Santa Barbara and elsewhere, who are least able to cope with climate change that extraction exacerbates; and the young, who will deal with climate change our entire lives. None of these communities were adequately represented at this meeting. With reference to the Latinx community, grassroots activist Arlo reflected:

> [At the meeting,] it was pointed out that we did a really bad job of reaching out to the Latino community in Santa Barbara. [. . .] There was some discussion on this for a while, and I'm looking around the room and there is not one person who is Latino sitting in the room, and it's kind of like, OK, we saw the mistake and nothing has changed. We are still right where we started; we have horrible representation within the Latino community, horrible outreach to the Latino community, and you know, hopefully, some of the organizations that were at the meeting have taken steps to change that, but it very much felt like there weren't. So that was something that we did not do very well in the Measure P campaign.[8]

Like Arlo, other activists in 350 Santa Barbara came away from the experience with perspectives emphasizing the need to build relationships with social justice groups in the county and with the Latinx community. Largely because of a decrease in core members' availability because of family and work responsibilities, only a handful of individuals made concrete progress in this area in the following two years. They tried, multiple times, to organize a coalition meeting with representatives from the Latinx community. They spent a year talking about the importance of building a multiracial coalition first and then deciding what to work on together. In a sense, they felt the need to construct a new table, rather than inviting people to a table they had already created. Implementation of these ideas failed to get off the ground because of capacity and scheduling conflicts. In 2016, however, Standing Rock and the November election of Donald Trump as US president renewed energy and provided events for these relationships to begin and grow.

In the context of the megaload campaign, convening a table looked very different. In Idaho, the grassroots network The Rural People of Highway 12: Fighting Goliath started with Borg Hendrickson, Lin Laughy, and their neighbors. The network was supported by larger organizations who worked with them in a horizontal manner. Other groups organized simultaneously in different locations, all with their own approaches to challenging the megaloads.

In Moscow, Idaho, members of the environmental justice task force of the local Unitarian church convened their own type of table in fall 2011, inviting members from local groups to meet to talk about their concerns and how they could collaborate. Many of the core groups who fought the megaloads attended the meeting, where people talked for hours. Pat Rathmann remembered, "People that hadn't ever talked to each other were talking to each other, and it was so exciting." A new group, Palouse Environmental Sustainability Coalition (PESC), formed during the meeting to serve as a network. Pat explained, "Any member of a group that wants to participate in PESC can join PESC. So individual groups don't lose their identity, and all individual members of those groups don't have to be part of PESC. We are a very kind of a loose-knit organization." Rather than creating one organization to control the campaign, as occurred with Measure P, groups followed their own strategies. This approach created conflict—as when organizations working on legal and administrative angles asked the direct-action group Wild Idaho Rising Tide not to protest on Highway 12—but was, overall, quite effective at involving many different grassroots constituencies that maintained their autonomy.

Part of this horizontal framework emphasized relationships, drawing on each person or group's particular skills and recognizing that everyone has something to contribute. This characterized the megaload campaign and the initial stages of Measure P. Rebecca August, who, following Measure P, started her own group where she worked to build on group members' "special powers," explained how she had been inspired by the horizontal grassroots energy when Measure P began: "It's empowering to connect with people and realize that it's just really about doing it and caring about it. [. . .] They [the people she worked with to collect signatures] didn't know anything more about anything really. [. . .] Nobody has any special information or skills or whatever. I mean, everybody has skills, but they're—you know, it's not a special person that it takes [to be an activist]." Samantha Smith echoed this view while reflecting on the megaloads: "You can work together and accomplish more if you have people that have different experiences. [. . . It] is nothing but positive. You may not always agree, but [. . .] you get to see another side of something that [. . .] opens your eyes to other things."

Relationships, and therefore people's capacity to leverage each other's skills, multiply when campaigns are organized horizontally. In Santa Barbara during Measure P, the Water Guardians employed what Han (2014) calls an "organizer model," in which they prioritized leadership empowerment among many people. Leadership happened at different levels. For example, core members of 350 Santa Barbara took on "team leader" positions, where they coordinated signature collection for an area of the county. This action was above and beyond the responsibilities that many had taken on up to that point. On the other hand, people who might just come to events or read newsletters took on the responsibility of collecting signatures, perhaps a type of leadership in the eyes of their friends and family members who were not involved in the initiative. In this type of organizing, relationship building comes to the fore as a group goal.

The Water Guardians were keen to build relationships for a variety of reasons. These included people power, funding, media, and political and community leverage. While a few people had been engaged in this type of work all along, during the signature-gathering phase, this work was more distributed than ever because of the nature of signature gathering. I, for example, went to more community events and built more relationships than I ever had before. These relationships paid off, literally and figuratively. One couple I connected with at a dance event that I attended for the purpose of collecting signatures became substantial financial backers of the campaign and future efforts.

Hazel Davalos, organizing director for Central Coast United for a Sustainable Economy (CAUSE), also prioritized movement building through distributed leadership development, which she had learned from the United Farm Workers movement. Her organization worked to inspire people to host house meetings and invite their neighbors. They would then all discuss their concerns and develop a plan of action. Hazel found this to be a very effective model among working-class Latinx people in northern Santa Barbara County, which suggests that Measure P might have been more successful if it had continued its horizontal organizing style for the entire campaign.

A final core component of effective relational organizing that interviewees highlighted was trust. This could be on an individual and organizational level. On the individual level, Gary Macfarlane explained, building a sense of trust and community is critical to motivating people and working together. Julian Matthews agreed: "I think realistically building the trust, getting the support of the people, the group of people, like the tribal membership, that's really the important point of doing anything." The Nez Perce willingness to trust white allies during the megaload blockade in August 2013 was vital to megaload protesters' fond memories of the campaign and to their continuing relationships. As Leontina Hormel explained, "The way Nez Perce people are treated in the Clearwater basin by non-tribal on average, certainly for them to trust allies is a huge gesture that in and of itself is pretty profound." Trust is also necessary to ensure that no one micromanages anyone else. With trust, Brett Haverstick thought, "there tends to become less of a need to micromanage because you learn to let certain parts of the campaign go, because you trust that individuals are going to do the work that the coalition is striving towards." This type of trust was very difficult to foster in the hierarchically organized campaign structure of Measure P, where a few people made decisions. Phone-banking scripts and preplanned neighborhood walking packets (used in Measure P) are ways in which traditional political campaigns efficiently micromanage all of their volunteers.

On the organizational level, trust increased the power of organizations. In Santa Barbara County, during election season, CAUSE appealed to the trust it built year-round in the community. It was not, in Hazel Davalos's words, an organization or campaign that "parachutes" into a community during elections. Instead, CAUSE worked each year to educate all voters, especially those who had not voted before, and to engage in community development. This approach, explained Hazel, was very different than most political campaigns, which tend to focus on "low-hanging fruit, the people who are the most frequent voters." "You can't fault them for that," said Hazel, "I mean that's strategic; it's like you have limited resources, and limited time. [. . .] But that's when Spanish speakers, low-frequency voters, newly registered voters, are getting left by the wayside."

CAUSE's nonprofit and political action designations—501(c)(3) and 501(c)(4)—allowed it to "do year-long work and develop relationships and accountability within the community" that then transitioned into support for political candidates or ballot measures. Hazel explained how this benefited CAUSE's ability to influence the community:

> We can say, "Hey, I am the person that fought to get that bus in your neighborhood last year. We are supporting this measure because it will be best; it is the best for our neighborhood," and that can go a long way, particularly I think in today's age of distrust of government. [. . .] If a community organization that has a track record of doing good work in the community tells you to vote for somebody, it's a little bit more effective [than being told who or what to vote for by a political party or candidate]."

The lack of trust in political parties and government that Hazel alludes to parallels the sentiments of members of CAIA in southwest Idaho, who avoided political labels like Democrat or Republican. In these cases and beyond, interviewees felt that organizations that were accountable to the communities they served, for the long term, would be most able to secure benefits for the community.

Hazel's comments also illustrate how relational organizing, rather than organizing based on conventional understandings of what is politically strategic or expedient, was necessary for creating inclusive and diverse politics, for getting new or marginalized people to engage—the very people who are most likely to support progressive politics.[9] CAUSE's success in this regard is evidenced by the fact that the organization's executive director and entire ten-person staff at the time of our interview were people of color. It also has a strong youth presence in its work, organized in the youth committee, which has spearheaded a number of successful local campaigns.

So far, I have presented horizontal and relational organizing as important elements of successful campaigns. Convening tables composed of stakeholders with varying levels of power, and creating campaigns where many individuals and organizations could lead, was a goal of the majority of interviewees. Building long-term trust and relationships were some of the most effective ways to accomplish this goal. In the following section, I examine commitment to diversity and inclusivity as another core component of building powerful social movements.

Commitment to Diversity and Inclusivity

Most grassroots interviewees prioritized diversity and inclusivity and felt that it was an important feature of resistance to extreme energy extraction. Motivations, definitions of success, tactics, and participant identities were facets of diversity that groups engaged with and tried to enhance. Motivations ranged from environmental values, quality of life, safety, climate change, climate justice, integrity, and accountability, as well as, most commonly, an underlying sense of injustice. Success for some meant legal and political victories, while for others, staying true to climate justice ideals, inspiring new supporters, and building grassroots power were all "successes." Various tactics included theater, occupation of space, letters to editors, Facebook page managing, court cases, ballot measures, house meetings, film screenings, and many more. The majority of interviewees, especially those not employed by an environmental or political organization, continually stressed that all forms of engagement make a difference and should be respected.

A few identified direct action as a tactic that is particularly undervalued. In a comment that clarifies why different approaches should be respected—and that relates to the Santa Barbara Water Guardians' decision to forego heavy investment in coalition building before beginning Measure P because of the urgency of climate change—Cass Davis defended the tactics Earth First! employed. Earth First! is a direct-action environmental group known for monkey wrenching, a term popularized by Edward Abbey's 1975 novel, *The Monkey Wrench Gang*, which refers to actions that destroy machinery used for environmental destruction (e.g., logging and mining equipment). Cass did not participate in monkey wrenching, but he was involved with Earth First!'s Cove Mallard Campaign, which took place in Idaho forests in the 1990s. One of Cass's roles was to try to explain to loggers why Earth First!ers engaged in direct action. To do so, he appealed to the sense of "emergency" as something that justified direct action. Cass explained, "[Earth First!ers] are fighting for life. They ain't just out there to screw up your bulldozer for something to do; they actually believe what that bulldozer is doing is destroying that which sustains life itself on this planet." Many interviewees had arrived at this sense of emergency in their ideas and emotions about climate change and energy extraction. In 2015, Cass felt like his role as an activist was to prepare a planetary hospice for people as climate and social injustices worsened. Helen Yost, Alma Hasse, and Becca Claassen were other interviewees who had reached high levels of emergency, which gave them a heightened level of energy and willingness to sacrifice time with family and personal comforts for activism. Respecting people's tactical decisions in the context of emergency was something with which they and many other interviewees could identify.

Respecting tactical choices made within certain political contexts was important too. Idaho activists were continuously aware of how their methods of organizing had to be different than those employed by organizations based in urban areas (e.g., Portland and Seattle) that they worked in coalition with. Helen explained, "Idaho is a hard place to work; I am freaking exhausted. [. . .] Some of our colleagues [on the West Coast] don't like us because we do things different. We have to because it is freaking redneck Idaho, and on some level, we are recruiting a lot more conservative people than they will ever have to recruit."

Therefore, "there's no silver/magic bullet," "you've got to use all of the tools in the toolbox," and "there's room and a need for many models" were common phrases. These statements referred to the type, framing, and scale of actions. Sharon Cousins's and Paulette Smith's perspectives exemplify these points. As Sharon put it, "Even if you can't stop [megaloads], delay them; it's all good, period. I never try and put down any methods of protesting. [. . .] In a suit in front of the White House, that's good. On the train tracks, that's good. I never want to alienate potential allies by going, 'Oh my way's better than your way.' We have different skills and different talents." Paulette pointed to a general opposition to the status quo: "I think in reality, our people [the Nez Perce] are waking up, I see more and more people standing up for things, and maybe it's not the same thing that I would stand up for, but dang it, they are fighting, and that's what I always hope for. Light a fire. I don't care what you're fighting for, but fight." Most interviewees aligned with Paulette and Sharon's view. They felt that diversity in the areas of motivations, tactics, and definitions of success increases the power of movements and their capacity to secure multiple forms of success.

Collaboration with Latinx and Native Communities

While Measure P and the megaload resistance were relatively successful in attracting supporters who were diverse in terms of gender, education, and class, they diverged in their ability to attract members of racial and ethnic minorities and Native nations. In the megaload case, collaboration with the Nez Perce Tribe was new and effective. It also was something whose strength grew as time went on. In the Measure P case, collaboration with the Latinx community, which made up a much larger segment of the population in comparison to the Nez Perce in Idaho, was not prioritized and not effective. In the aftermath of the measure, there was little evidence of action on the part of grassroots or grasstops groups to change this situation. Despite the good intentions and actions of some individuals, a lack of clear progress by entire groups meant that relationships were not built into the structures of organizations, which is critical to sustaining relationships for the long term.

The two quotes that follow, one from Dave Davis, the executive director of the Santa Barbara Community Environmental Coalition during Measure P, and the other from Brett Haverstick, one of three staff for Friends of the Clearwater in Moscow, Idaho, illustrate the contrast in sentiments about capacity for multiracial and multination organizing in these cases.

> Environmental communities do *not* connect to the Latino or Black communities in the south coast [of Santa Barbara County]. [. . .] We talk about [diversity] all the time. How do you connect with the community? I don't find good connections [between the environmental and Latino community].
> —Dave Davis

> In looking at the megaloads, the inclusivity was really unique or special in the fact that it really strengthened relations between Indians and non-Indians. [. . .] At the end of the day [. . .] you need racial diversity [. . .] you need youth, you need adults. [. . .] When people hear a message coming from

a five-year-old girl [. . .] or a twenty-five-year-old Nimiipuu or Nez Perce tribal member or a ninety-five-year-old Caucasian, it starts to resonate with people. [. . .] People start realizing, "Well heck, there's something really big going on here; this is affecting a lot of people, and I'm going to look at it, you know, because of that."—Brett Haverstick

The success and value that megaload protesters placed on diversity of participants was markedly different from the Measure P experience.

In Santa Barbara, how to attract people of color to work on environmental and climate issues was, for most interviewees, a challenge they identified as a weakness in the environmental community, which is, in Daraka Larimore-Hall's words, "super-white." Dave Davis, who had lived and organized in Santa Barbara for decades, said that in the seventies and eighties, things were different. There was a lot more collaboration between different racial and ethnic communities in Santa Barbara, something he attributed to personal relationships. These relationships had not been sustained in organizations, leading to the lack of collaboration in the present era that he and so many other interviewees lamented.

In Santa Barbara, interviewees highlighted three main reasons for the lack of collaboration between the environmental and Latinx communities: lack of personal connections; lack of immediacy of environmental issues, compared with issues like rent, immigration, and police interactions; and insufficient broad-based grassroots organizing within the Latinx community. Many of these same issues likely informed why there was not large involvement of Chumash tribal members in the Measure P campaign. A few Chumash individuals engaged and were invited to participate, performing opening ceremonies at fundraisers and events, but no Chumash people or organizations were among the core organizers or group endorsers of the campaign.

Spatial segregation between white and Latinx residents contributed to lack of personal connections, and indeed, during the years leading up to and following Measure P, the local Latinx community had focused efforts on wages, rent control, farmworker rights, a gang injunction that targeted Latinx youth, and resources for undocumented communities. In Lompoc, Latinx community members collaborated with Measure P activist Janet Blevins and the local Democratic Party to help undocumented people obtain drivers' licenses.

In terms of Latinx community organizing, Daraka Larimore-Hall perceived a lack of stable organizations with broad-based Latinx support to collaborate with in South Santa Barbara County. Like other interviewees who had been in Santa Barbara for some time, he lamented the weakening of a group called PUEBLO (People United for Economic Justice Building Leadership through Organizing), which had been very strong within the Latinx community in Santa Barbara. PUEBLO's formation had followed a period when the Santa Barbara Latinx community, particularly the working class, was "under-organized and under-mobilized politically for a long time," according to Daraka. PUEBLO, however, collapsed and was then absorbed by the regional group CAUSE in 2012. During this period, what Daraka called a "vacuum," two kinds of politics emerged in the Latinx community in South County. One was "business oriented" and conservative. It was grounded in a group called the Milpas Community Association (MCA), which worked to install a business improvement district (BID) in east Santa Barbara that opponents perceived as a gentrification effort. The other, a group called People Organizing for the Defense of Equal Rights (PODER), was more radical and militant. "There was nothing in between that was sort of like organizing based, pragmatic, good analysis of power, that PUEBLO was doing and trying to do," explained Daraka. The gulf between these two wings of the local Latinx community and a more centrist approach, and especially the gulf between South and North County Latinx organizations, is evident in the CAUSE community organizing director, Hazel Davalos's response to a question I asked during our interview:

Corrie: Do you ever work with PODER?

Hazel: Sometimes. [. . .] We walk an interesting line as an organization; I think particularly out here [Santa Maria] in conservative circles, we are seen as very radical to the left, and by some organizations [likely referring to PODER], we are seen as too mainstream, too moderate, because we actually tell people to vote for candidates. Some people, who have a school of thought that all politicians are corrupt [. . .] are critical of that piece of our work. So, we are not able to work with every single group.

The disconnect between Santa Barbara city Latinx groups and the environmental community was made clear during Measure P when a Latina resident of Santa Barbara, Jacqueline Inda, who had taken many of the same stances on local issues as PODER, took money from oil companies to be a spokesperson for No on Measure P. She appeared in No on P ads spreading the message that Measure P would hurt Latinx families. Daraka explained, "The fact that there is no good antiracist social justice leadership in the local environmental community is one of the factors that allowed that kind of thing to happen. [. . .] I'm not singling out the environmental movement as the only culprit there. [. . .] Getting to this bad situation had a lot of authors, a lot of authors, but it was one of the weird ways that race played a role in the Measure P fight."

To summarize, there was a lack of communication and relationships between the Latinx and environmental community in Santa Barbara. Combined with splits within both of these communities, this lack of connection exacerbated barriers like language and clarifying links between social justice and climate change—barriers that individual activists struggle with. These divides were clear in the election results, where predominantly Latinx North County communities voted overwhelmingly in opposition to Measure P.[10]

Within this context, Santa Barbara interviewees had a number of ideas on how to improve diversity. These proposals included being very active as an organization to provide events that people could join in on (Katie Davis), building coalitions with social justice groups (Alex Favacho), and building personal relationships of solidarity by going to events or group meetings targeted to racial justice, immigration, and other causes organized by people of color (Max Golding). The majority of interviewees suggested ramping up outreach to the Latinx community in culturally relevant ways, on popular media platforms, and in Spanish. The inadequacy of the Spanish-language outreach on Measure P is illustrated by the fact that I translated the bulk of "Yes on P" written materials into Spanish. I am a non-native Spanish speaker with no experience speaking Mexican or Chicanx Spanish. A local Peruvian climate activist and a few of my graduate student friends looked over the materials, but there were no local Spanish speakers who played a lead role in the process. A final idea, offered by Hazel Davalos, whose entire organization was mostly composed of working-class people of color, was that white allies hold a special meeting to invite representatives from other groups in to explain their work and hear their ideas.

In Santa Barbara County, models along the lines of Hazel's and Alex's suggestions, when combined with good outreach, are likely to be the most effective. As I explain in the coming pages, these were the methods that worked well in the megaload case. The strategy of climate and racial justice activists building a table together is something that many interviewees wanted to do. In practice, however, few had made progress in this area. Max, who was a proponent of this idea, was cynical about its feasibility: "You just can't, like you can read about [. . . diversity], but [. . .] if nobody looks like you within the people that you are [wanting to join]—it's like none of us [350 Santa Barbara] have ever gone to Latino groups; like why would we? It would be really awkward, because there's this cultural bridge. Like we are all open-minded enough to where we would be open to it in theory, but none of us ever have or ever will." Little effort was made by 350 Santa Barbara and the Water Guardians before or after Measure P to attend Latinx group meetings or to coordinate a cohesive group presence at Latinx events (again, individual climate activists attended these events, and sometimes brought group signs, but did not typically attend as a climate/environmental organization). Max and others recognized the reality of low group capacity as a factor in why 350 Santa Barbara did not build a multiracial coalition but did not see this as an excuse.

This lack of action to build a diverse environmental movement is even starker when examining the 2017 leadership of the Santa Barbara grasstops groups (see table 2), which have significantly more resources than 350 Santa Barbara. Also, note the predominance of women as staff and volunteers in these organizations, something that is generally consistent throughout my research.

To put these numbers in context, the US Census Bureau (2019a) estimates that 46 percent of the residents of Santa Barbara County are "Hispanic or Latino," compared with 37.1 percent of residents of the city of Santa Barbara. Only 43.8 percent identified as white alone, not Hispanic or Latino, in Santa Barbara County, and only 55.6 percent identified this way in the city of Santa Barbara proper. All groups in the table who are based in the city of Santa Barbara fall woefully short of approaching adequate representation of people of color. I bold the numbers for CAUSE in the table to draw attention to just how different they are from the environmental groups. This contrast is due in part to the fact that much of their Santa Barbara County organizing occurs in Santa Maria, which is 76 percent Latino, on issues such as farmworker rights and immigration—issues that are immediately salient to this community, part of everyday life. As my analysis suggests, however, their organizing style also plays a role.

Turning to Idaho, there is sharp contrast in the commitment to diversity evidenced by the leadership of Friends of the Clearwater, the environmental organization with paid staff that was most active on the megaload issue (see table 3). While the numbers are small, the proportion of Nez Perce represented on the board of directors of Friends of the Clearwater is much greater than in the general population. In 2010,

TABLE 2 MEASURE P: DEMOGRAPHIC CHARACTERISTICS OF LEADERSHIP OF
LOCAL ENVIRONMENTAL GROUPS AND CAUSE, FEBRUARY 1, 2017

Organization	Total	Men	Women	Potential Person of Color[a]
Community Environmental Council (Board of Directors, officers, and members)—all former leaders of the group were white	10	7	3	1 (10%)
Community Environmental Council (staff)	9	2	7	2[b] (22%)
Environmental Defense Center (Board of Directors)	15	9	6	0 (0%)
Environmental Defense Center (staff)	9	2	7	1 (11%)
Santa Barbara Sierra Club Group (Volunteer leadership in 2016)[d]	10	5	5	2[c] (20%)
Los Padres Group of the Sierra Club, Santa Barbara and Ventura Counties (Candidates for volunteer leaders in 2017)	8	1	7	0 (0%)
CAUSE Board of Directors	14	8	6	6 (43%)
CAUSE Staff	11	4	7	10 (91%)

a. This is based on a person's photo and name. While it is an inadequate way to measure race or ethnicity, looking at a person's photo and their name on a website is what a potential new member of the organization would likely do. In reference to Max Golding's quote, it provides a useful proxy for thinking about who would and would not see people who look like them in an organization.

b. One is from Malta.

c. One I know to be biracial.

d. Note that this is a grassroots group, but it has a much longer history than 350 Santa Barbara.

in Latah County, Idaho, where Friends of the Clearwater is based, 90.6 percent of people identified as only white, with about a 1 percent estimated decrease in 2015. And even in Nez Perce County, where towns within the Nez Perce Reservation are located, 88.7 percent of residents identified as white alone, neither Hispanic nor Latino, in 2010. Just over 5.5 percent identified as American Indian alone. Another important point is that two of the Nez Perce individuals on Friends of the Clearwater's board of directors joined after most of the megaload actions were over. This demonstrates a commitment to improve the representation of diverse voices in environmental issues, based on what environmentalists learned from the megaload struggle.

TABLE 3   MEGALOADS: DEMOGRAPHIC CHARACTERISTICS OF LEADERSHIP OF
LOCAL ENVIRONMENTAL GROUP AND NIMIIPUU PROTECTING THE ENVIRONMENT,
FEBRUARY 1, 2017

Organization	Total	Men	Women	Person of Color
Friends of the Clearwater (Board of Directors)	9	7	2	3[a] (33%)
Friends of the Clearwater (staff)	3	2	1	0 (0%)
Nimiipuu Protecting the Environment (Board of Directors)	4	2	2	3 (75%)

a. Persons of color are the same three Nez Perce individuals in both organizations. This puts much of the burden for Native–non-Native organizing on the shoulders of these individuals. As Nimiipuu Protecting the Environment grows, hopefully other individuals can take on some of the labor of uplifting Nez Perce interests in regional organizations.

Further evidence of the progress made during the megaload campaign, in terms of working across Native and non-Native lines, comes from interviews with Nez Perce women, who stressed how good it was to see non-Native allies at their megaload blockade. Chumash individuals expressed similar views during Standing Rock protests held in Santa Barbara in 2016, hinting at potential for change in Santa Barbara County as well. Nez Perce tribal members Paulette Smith, Lucinda Simpson, and Samantha Smith expressed their views in the context of a group interview where we used the method of a listening circle. Using my recorder as a talking stick, we passed it among the four of us, listening as each person gave her account of the theme under discussion. In keeping with this method, I highlight their three perspectives below. In Paulette Smith's words:

> I think that [networking] is the only way we are going to be able to become this united huge front—that's my dream, is that we are all across the nation, different colors and races. [. . .] I was totally impressed during the megaloads, with all the non-Natives that showed up, and I'm like, yeah, we need that. [. . .] We have to become this force; we have to stretch across wherever and however far we can. [. . .] We've gotta agree to disagree, because I don't agree with all the people we work with, but I think the big picture—we're all working toward that. So that's how I see it, is that network.

Paulette echoes the "agree to disagree" mantra of activists in southwest Idaho and, like them, stresses the importance of finding a common goal to bring people together.

Lucinda Simpson, a tribal elder and board member of Nimiipuu Protecting the Environment, who had recently joined the board of Friends of the Clearwater, also prioritized coming together:

> The more people that you have, the better things work out for you, the more minds you have in one room that have different ideas. I think that if they all come to the table with one common goal, I think that's a big asset for us. I think that it's taken them [environmental groups] a little while to get used to us. They've known Julian Matthews for some time, but they are adjusting to us women [laughs] and getting an understanding of us a little bit too. I've been to different meetings up in Moscow with the Friends of the Clearwater [. . .] and I've had my chance to speak the way I feel about different things.

Nez Perce tribal member Julian Matthews had long been in touch with environmentalists in the area and had started Nimiipuu Protecting the Environment in the wake of the megaloads. His personal relationships with Native and non-Native individuals had facilitated relationships among people from these communities, evidenced by Lucinda's excerpt.

Finally, Samantha Smith explained what it had been like to grow up in Idaho as a Nez Perce woman. Twenty-six years old, she had experienced intense grief, stress, and anger upon hearing that her mother, Paulette, and her sister had been "manhandled" during one of the megaload blockades while Samantha was out of town and unable to help. Experiences like this, which resonated with a whole history of trauma for her people, made the fact that white people stood in solidarity with the Nez Perce a surprise for Samantha. She explained,

> I guess growing up around the area, you encounter a lot of different people, some of them really nice, some of them not so much. So it's always kind of a shock to see non-Natives turnout for something like that [the megaload blockade]. [. . .] You don't expect non-Natives to come to your rescue, especially in this area, and I mean, it's a big help to minorities to have the majority there. [. . .] It speaks volumes; it means that there are humans in the world that care. [. . .] Finding those people that just see you as people and are willing to stand alongside you is amazing, because it doesn't come along a lot. So, I think the fact that this group [Nimiipuu Protecting the Environment] is working with others that have different life experiences, different levels of privilege, it is a huge help.

The megaload campaign brought people together across Native and non-Native lines. It illustrates how organizing against a shared enemy can be a springboard for talking across lines, identifying common values, and building new campaigns together.

Multigenerational Collaboration

A final component of diversity for which interviews offer insight is multigenerational collaboration. Age arose organically in many interviews as an area of tension related to building inclusive movements. In Measure P and the megaloads, a wide range of age groups participated. While megaload interviewees were unanimous and explicit about the importance of involving young people, there were differences in how explicitly in support of young people Measure P proponents were.

With the megaloads, Brett Haverstick talked about involving elementary school children, showing me drawings that local students created after he gave a presentation on the megaloads at their school. One of the opening quotes to the megaload section of the previous chapter comes from one of these students. Similarly, Paulette Smith and Julian Matthews emphasized the need to involve Nez Perce youth in community organizing and environmental issues, as well as the need for others to learn communication strategies involved in social media so popular among youth.

In contrast, many of the youth who participated in Measure P felt alienated by the mostly old, white, and wealthy members of Santa Barbara's environmental groups. The leaders of these groups were also mostly men. Non-youth members of 350 Santa Barbara perceived a lack of support for youth engagement as well. Katie Davis, age forty-five, explained, "It was surprising to me. [. . .] There was some like negative, reaction [. . . to] these young people coming out and doing stuff. [. . .] There was a little weird territorialism amongst the old [environmental community in Santa Barbara]." Gary Paudler, age fifty-eight, was disgusted by established organizations who "refused to be challenged by an upstart organization, and especially one with a lot of young people." Dave Davis, age sixty-seven, expressed the opposite view, likely resonating with the old guard environmentalists, who were also older in age. In reflecting on Measure P, he said: "I say this with all due respect, it was like you guys [Dave and others] are old guys, and you guys just don't have the passion anymore [. . .] and we were turning around and saying you guys [350 Santa Barbara] are young guys, and you don't realize what you are doing. [. . .] So, we were talking past each other, and unfortunately, we were talking past each other very late in the game." Coming together across lines of age—a proxy for presumed life experience relevant to political organizing—posed challenges in Measure P in Santa Barbara and among environmentalists working on other, non-megaload issues in Idaho. Young and old activists felt alienated by the other group, inhibiting each group from learning from and teaching each other. Kelsey, former executive director for an environmental nonprofit in Idaho, age thirty-four, offered a helpful metaphor for bridging this divide. Rather than having older activists "pass the torch" to younger activists, she advocated for "lighting a new torch." Kelsey explained:

> The idea is that you light another torch [. . .] and you both have light, and there is more light. That's what I would like to see happen more in this intergenerational transition, is they [older people] can stay involved and give us their wisdom and learn from our passion and our new ideas, and we can temper our passion and new ideas with the wisdom and the experience, and we can find some way to actually solve these problems—because the problems are still happening, and it's like, yeah, I get that you started your program in the '80s, but you haven't fixed it yet, so maybe our ideas could help.

In the relatively successful example of working across lines of indigeneity and age—the megaloads—these identities are clearly intertwined within particular contexts. Like other Indigenous communities, the Nez Perce hold elders in high regard. They also think far into the future, in terms of seven generations. Some of their cultural practices around age, therefore, likely shape interactions with non-Native activists and the character of joint projects these groups will take on in the future.

CONCLUSION

The Idaho megaload and Santa Barbara Measure P campaigns took two different approaches to resisting extreme energy extraction. The megaload effort had multiple levels of engagement, in line with participants' preferences. It drew on diverse concerns, appealed to people's connection to place, and articulated a unified stance against all elements of the megaloads grounded in a sense of injustice, made more possible in a non-extraction state. After qualifying for the ballot, Measure P became an effort with few avenues for grassroots engagement, against the wishes of core participants. Reacting to industry messages emphasizing Santa Barbara's dependence on oil obscured proponents' messages. They were unable to convince enough voters of the injustice of extreme energy extraction.

These cases illuminate effective and ineffective ways to construct diverse coalitions with enough power to accomplish their goals. Connection to place, political economy, and identity were each important variables in these struggles. The comparison of these cases suggests that organic mobilization among diverse constituencies, when welcomed and included by others, can build foundations for working across lines. Valuing relationships, prioritizing trust, and recognizing shared values all support this type of outcome. Showing solidarity, improving representation of people of color in the leadership of existing organizations, and valuing the knowledge of new and old generations are concrete paths forward.

This chapter draws on and extends the analyses in previous chapters by demonstrating how methods for building climate justice cultures of creation, talking across lines, and collaborating across organizational forms resonate widely among interviewees. In many ways, it analyzes large-scale efforts to implement the values youth interviewees espouse— the values at the heart of their political culture—among a set of activists who are much more diverse in age, experience, class, and education. I argue that these methods for working across lines are the connecting threads among concerned community members protecting property in southwest Idaho, college students organizing in Santa Barbara, representatives of grassroots and grasstops organizations, and the diverse individuals who animated the Measure P and megaload struggles. These methods are one unifying characteristic of the climate justice movement.

Conclusion

I began this research with a contradiction: as society is becoming ever more aware of the risks of climate change, it is simultaneously expanding extraction of fossil fuels, and not just any fossil fuels, but the dirtiest, most expensive, and hardest-to-reach types. In this context, I felt that the logical response was resistance. I joined in with resistance in Santa Barbara, California. For a time, I found it invigorating. It made me feel at home in a place and connected to people with whom I enjoyed spending time—it gave me a sense of community.

I began to be curious about why organizing resistance gave me, and many of the people around me, enough joy to put in the work that organizing requires, on top of schedules already full of work and family obligations. I wanted to explore other people's experiences—to explore the nature of resistance to extreme energy extraction in different places. I thought that better understanding the contours of resistance might be useful for strengthening that resistance—for illuminating the creative ways of resisting that different folks have developed to meet the challenges and opportunities they face. I spoke and organized with many different people in communities in Idaho and California. Despite living in two very different contexts, their ideas about resistance returned, again and again, to ideas about how best to work together with other people. More specifically, the people I talked with were concerned about how to work together with people who had different views, different identities, different priorities, and different organizational affiliations. They wanted to be good at working across lines.

In my research cases, I argue that working across lines undergirds resistance to extreme energy extraction. Drawing on ethnographic fieldwork conducted throughout 2013 to 2016 and 106 in-depth interviews, I find rural Idahoans working across lines of political identity in southwest Idaho and college student activists working to create inclusive climate justice organizations with diverse participants and campaigns. I find grassroots and grasstops groups in tension, with members of both types of organization seeking methods for working together. Finally, through exploration of two completed campaigns, I find evidence that building a broad-based network that welcomes rather than attempts to homogenize different motivations, tactics, and strategies can be key to success. Within the paradox of extreme energy extraction and climate crisis, people are working together to envision and embody a society that is just and sustainable—in other words, to build climate justice.

Interviewees' actions are building the climate justice movement, both by inspiring people to become politically engaged and by practicing, sharing, and developing ways of being with other people and the planet that will be central to achieving climate justice. They are working across lines by:

- focusing on core values, which include community, justice, integrity, accountability, and health of people and the more-than-human world;
- identifying the roots of injustice, whether described as capitalism or lack of integrity and accountability of government and industry;
- cultivating relationships, what interviewees call *relational organizing*; and
- welcoming difference.

In southwest Idaho, working across lines takes the form of talking across lines, where activists focus on messaging issues that resonate widely in their community, avoid reinscribing each other into stereotypes related to political party affiliation and climate change beliefs, and agree to disagree on some topics so they can come together on others. Over time, and with relationships of trust, I think even topics that activists agree to disagree on can eventually become subjects ripe for collaboration or, at the least, respectful conversation. In college organizing environments in Santa Barbara, California, working across lines is a creative culture of values and practices for building relationships, making organizing accessible, recognizing the intersectional nature of people's identities and social movements themselves, and prefiguring communities where people are politically engaged and recognize their codependence with others, even those they do not agree with.

Within the tensions that characterize grassroots/grasstops collaborations in my research, working across lines will be facilitated by more willingness, on the part of grasstops, to welcome new ideas and center, rather than marginalize, solutions that address the root causes of climate change, even if they are not pragmatic within the current political economy—a capitalism dominated by inequality and powerful corporations. Collaboration will also be facilitated by all types of organizations and activists prioritizing the construction of coalitions that include all sectors of the community. Working across lines can be very difficult in short time frames, when opposed by the oil and gas industry's financial power, and in hierarchically structured electoral campaigns, as I show with Measure P in Santa Barbara County. In contrast, a loose network that allows each group to use their preferred tactics and strategies can inspire many different people with different motivations to support a struggle. As the Idaho mobilization against megaloads demonstrates, a healthy coalition can cultivate enough people power to foil the plans of a modern-day Goliath, ExxonMobil. That rural white Idahoans could work together with the Nez Perce Tribe in one of the most conservative states in the country to bar ExxonMobil and other companies from the most cost-effective route for transporting infrastructure to the tar sands should be a story of hope for activists everywhere. It is an example of how relationships, trust, solidarity, and coming together around shared understanding of right and wrong enables working across lines.

Working across lines is both a practice and vision that can be what Tsing (2004) calls a "charismatic package," a traveling activist model that can feature morals, stories, or organizational plans. These packages can show how the underdog can become a political force; they are informed and given meaning and power by location and context, which can be a place, widespread political culture, or point where two cultures meet (Tsing 2004:227–28). As Tsing writes, an activist package "is most striking when it inspires unexpected social collaborations, which realign the social field. We might speak of 'happy collaborations'—not joyful but felicitous [fitting]; in coming together, they make a difference in what counts as politics" (227–28). Working across lines is an

FIGURE 12. Students race on carbon bubbles, rubber balls that say "gas," "coal," and "oil." Photo by the author.

activist package with these goals at its core. And, I would argue against Tsing, it is an activist package that strives to be joyful. This is evident in the creative protest tactics employed by interviewees in many locations—students participating in the May 2016 Fossil Free UCSB Divestival hopping on carbon bubble balls or pinning the lie on Exxon (see figures 12 and 13), women dressing up in ball gowns to stop megaloads, and CAIA members erupting in laughter (after a six-hour evening meeting) while brainstorming a holiday fundraising item, a large button that when pressed would say, "You are *so* horribly fracked!!" Joy is central to persevering (see Bell 2013:39) and prefiguring the world activists want to see.

Working across lines through the four methods I describe is also a way to build a collective identity. Scholars define collective identity as activists' shared understanding of the problems they face, the solutions they propose, and the environment in which they organize, as well as the relationships they build around these understandings. A collective identity is a process rather than a thing. It is continually negotiated and reshaped, through activists' relationships with each other and the emotions that infuse a social movement (Martínez 2018; Melucci 1989; Taylor and Whittier 1992).

FIGURE 13. In "Pin the Lie on Exxon," student attempts to pin the lie on then Exxon CEO Rex Tillerson's face. Photo by the author.

Collective identity is a particularly important part of organizing for the climate justice movement. As with other movements rooted in justice, the climate justice movement's goal is for those most affected by the problem, the people on the "frontlines," to be at the forefront of solutions. The logic behind this approach is that the people who experience the worst effects of a situation are most well equipped to identify solutions and that the solutions that help the most marginalized will benefit everyone. In her study of the barriers to mobilization against mountaintop-removal coal mining in central Appalachia, Shannon Bell (2016) explores collective identity in the environmental justice movement. She finds that what people on the frontlines of mountaintop-removal coal mining perceive as the collective identity of the environmental justice movement dissuades them from participating. This is a major problem for a movement whose central tenet is leadership by the people most affected by environmental injustice.

I argue that the methods for working across lines that I detail in this book possess potential to address this problem. They can facilitate activists' capacity to build collective identities that are inclusive to many different communities, both because they are grounded in cultivating relationships, seeking to clarify how social and environmental issues are intertwined, and because they strive to prioritize messages that resonate with people across political affiliations, identities, and positionalities within organizations. In these respects, working across lines can illuminate methods for building collective identity, coalitions, and, more broadly, political cultures of resistance and creation (Foran 2014; Reed and Foran 2002) in many different social movement contexts.

This research also builds understanding of how people engage with their environment, which is the core question of the environmental social sciences. The social drivers and consequences of the climate crisis, and the marginalization of people-centered perspectives on how to address this problem (Dunlap and Brulle 2015; Sovacool 2014) reinvigorate the urgency of expanding knowledge on the current state of this engagement and the innovations that people are developing. I explore the climate justice movement as one of these innovations. As a relatively new movement, it joins other twenty-first-century movements in having the privilege of drawing on examples and insights from more historical movements and thinkers than ever before. Digital technologies make the scale and speed of sharing this information unprecedented. Understanding how this movement wields this privilege, and how it might do so in the most just, sustainable, and powerful way possible, enhances understanding of how social movements work and how they can transform socio-environmental relationships.

Transforming these relationships is critical and urgent. The contradiction of extreme extraction and climate change is unprecedented in its scale. It reveals, as never before, the "slow violence" (Nixon 2011) of status quo social relations—how, if high-carbon societies continue current practices, there will be untold global suffering and damage. We do not have to do anything (literally) to arrive at this dystopian future. It will affect all living beings' capacity to survive and has already begun. The climate emergency is harming many, while countless others as yet feel untouched by its effects. It is as if massive nuclear bombs have been launched globally, with some having already hit low-lying and marginalized communities. What remains to be seen is how we respond.

The increasing clarity of our circumstances, however, offers openings for change. Energy and climate affect the lives of everyone on earth in intimate ways. A Wall Street banker could not survive without energy, just as a Hadza hunter-gatherer in Tanzania could not survive in a climate drastically different from the one to which her skills are adapted.[1] Climate change threatens the resources that all people depend on. Extreme energy extraction is capitalism's last-ditch attempt to stabilize its crumbling structure, a structure that cannot work without inequality and ever-increasing use of resources. In this attempt, capitalism is, as Marx would say, digging its own grave, both materially and socially. As renewable energy becomes more affordable, as local climate impacts become more widespread, and as extreme energy infrastructures disrupt the lives of more people, maintaining the status quo seems unacceptable to an increasingly large portion of the population.

People respond to this realization in different ways. As the election of Donald Trump to the US presidency in 2016 illustrates, some people respond by demanding change, no matter how outlandish and offensive the form of change may seem to people who hold other views or who experience oppression because of that change. From my interactions with research participants, I think movements for justice can harness this desire for change by identifying common values, practicing deep listening, cultivating trust, and welcoming difference as the very core of what makes solidarity and collaboration such a rewarding social experience. With these methods, change can propel us forward to a world where people enjoy just and sustainable relationships with each other and the more-than-human world.

MOVING FORWARD

Moving forward, I hope this research will spark exploration of working across lines in other places, movements, and times. What are other examples of unlikely alliances? What methods have alliance participants used to work together and how have these contributed to achieving their visions of success? How do different cultural settings and histories inform who participates and how? How will dividing lines—whether political, racial, or cultural—shift as the climate crisis accelerates?

When possible, research on these themes is richest when conducted during a campaign. By rich, I mean able to generate emotion in the researcher and reader. Emotion is important for empathy and, in my experience, facilitates understanding of how other people's lives relate to one's own—in essence, a person's capacity for "sociological imagination" that connects personal troubles to social issues (Mills 1959).

I engaged in research during a campaign in southwest Idaho, where daily life for the people I lived with revolved around how to work across lines. As a scholar activist during this period, I had time to reflect on and synthesize the experience while it was fresh in my mind and most useful to activists. My research on Measure P blended contemporary and retrospective analyses. In contrast to my involvement in southwest Idaho as a scholar activist, during the Measure P campaign I played the part of an activist scholar. Because my scholar role was secondary, the bulk of my analysis of the event, and my efforts to capture participants' perspectives, occurred two years later. I engaged with the other campaigns that constitute the subject of this book in retrospective fashions. Retrospective accounts have the benefit of facilitating interviewees' capacities to express their views in distilled forms that grow out of the coalescence of lived experience, social context, and change over time. Yet these accounts cannot enable the researcher to witness how and whether interviewees act on their perspectives. Some of this witnessing can happen through reading news reports from the time, other researchers' accounts, or by watching videos that are posted by participants on the internet, which I have done. These mediated forms of experience facilitate different analyses that contribute to, but cannot replace, research conducted by someone with the time and privilege to engage in scholar-activism as working across lines unfolds. I feel fortunate to have had this privilege for portions of the research. Yet I am also grateful for my activist-centered experience of Measure P, which inspired this project. Future research should take advantage of the many struggles underway, documenting, analyzing, and synthesizing lessons learned in real time, for scholars, activists, and people living within a changing climate.

Some readers may wonder how the lessons from *Working across Lines* apply in a United States that is much more divided than when I conducted this research. In the post-2020 context shaped by uprisings for racial justice in response to the May 2020 police murder of George Floyd and so many other Black people; the politicization of the COVID-19 pandemic; and the January 2021 seizure of the US Capitol by then president Trump's supporters, acknowledging and working through difference, rather than ignoring it, is more urgent than ever. Building relationships is critical to building coalitions across difference. However, the kinds of coalitions organizers decide to build should be based on context and strategy. Racist groups or groups that exist to amplify national polarized ideas at the local level are likely not where climate justice activists should spend their energies. Real grassroots organizations, however, composed of local residents who may have different views, but are not actively causing harm to marginalized communities, may be useful allies in fighting extreme energy and building a healthy future.

When I am challenged by the labor involved in working across lines, I think of Indigenous activist Winona LaDuke, with whom I have been collaborating the last couple of years against the Line 3 tar sands pipeline in Minnesota, my new home. She always says that she wants water protectors to be perceived as "the home team" in Northern Minnesota (see LaDuke 2019), a place, like Idaho, that is generally conservative and where racism and discrimination against Native people is clear. If she, as a Native woman who has been doing environmental justice work for much longer than I have, still has hope of building relationships with non-Native conservative neighbors to support Indigenous water protectors, then I feel obligated to continue the effort to work across lines. As Rebecca Solnit writes in *Hope in the Dark*, "Hope locates itself in the premises that we don't know what will happen and that in the spaciousness of uncertainty is room to act" (2016:xiv). *Working across Lines* provides methods for acting, even amid the uncertainty caused by the multiple crises—of democracy, racism, public health, and climate change—that we all face. Among groups working for social justice, accountability, and integrity, the lessons from *Working across Lines* will be helpful in building broad-based and powerful movements capable of creating a thriving and equitable world.

In closing, I highlight interviewees' reflections on hope, something they locate within their collaborations with other people—while working across lines. Hope is critical to realizing climate justice now and in the future. In each of my 106 conversations with research participants, my final question was, "What do you hope for the future?" Here are a few of their responses.

I'm interested in and hopeful for a generation of our young people, your generation, to live well and thoughtfully of neighbors and carefully with creation and innovatively with technology, and more simply than our generation does. And what do I think about the prospects? [. . .] You know, I'm optimistic because I can see the narrowness, the death, if you will, in kind of where things are headed and I also see the beauty of the earth, the beauty of community relationships, the beauty of innovation, and the beauty of simplicity, and I just am not sure how people can't be attracted to that if they will open their eyes. [. . .] And if it doesn't happen, I still want to be optimistic. Because the alternative is just not joyful. [. . .] I would rather keep a level of optimism in people being able to receive truth than condemn people to stupidity and embracing falsehood. And there are a lot of really wonderful young people around who are looking for vision and looking for something to do with their life that will be joyful, and I'm all for them; I think they're going to do well. —Peter Dill, age sixty

We are either going to bring about [a] beautiful future, or we are going to be the hospice workers for the dying future, you know, and either way, that is noble work. [. . .] I'll dedicate my life to that. [. . .] We're going to put all of our love and all of our energy into it and just hope for the best. And I don't see complete tragedy in the future. [. . .] There are so many smart minds working on this, like we have hope, definitely. I'm kind of excited [laughs]. —Miranda O'Mahony, age nineteen

When I see something positive in the news or, you know, something like what happened in Portland, like people hanging off the bridge to block the [oil] ship, that just makes the cockles of my heart warm, and I feel like there's enough of us who care about the planet out there—it's just a matter of finding the means to make it happen and also organizing. [. . .] There's tons of us who want that to happen, so I just feel like it's a matter of time, us doing everything we can to actualize that.—Zach Rosenblatt, age twenty-three

As long as there is community there will be hope. There has to be; people are so powerful—they just have to realize it. A lot of people don't even realize how much power they have within themselves, so community is good for that, because when you have someone else telling you that you can do it, then you're like, "Yeah, I can do it." And that's what we need more of; we need more people empowering each other to do what they want to do.—Camile, age twenty-one

There are a lot of [. . .] people who [. . .] have found that they do have power [. . .] and they are using it, and they are getting more and more engaged in their region and in the community and making connections. [. . .] Knowing that these people are working hard, it gives me [. . .] a sense of hope that we are going to be OK.—Arlo, age twenty-three

My hope really is that we don't survive, but we thrive. [. . .] I really hope that people can come together and learn to work with each other and talk to each other, and we get out of this kind of the domineering mindset and really create healthy alternatives to capitalist social relations.—Jake, age twenty-four

Nothing's really going to change much unless everyone is able to see where everyone else is coming from and cooperate because the powers that be are very, very, very powerful. [. . .] It's going to take a lot of people being sensitive to one another to be able to confront that effectively.—Sarah, age twenty-three

With the exception of Peter, whose words are a beautiful example of an older generation's hope, these thoughts on hope all come from young people whose entire lives will be shaped by the climate crisis. That they have uplifted the importance of love, community, empowering each other, and cooperation bodes well for the future.

The creativity of young people, combined with the experience and willingness, on the part of elders, to welcome and nourish new ideas, and the empathy, from people resisting extreme energy extraction everywhere, to realize the shared roots of their struggles, will all be critical to working across lines. As Sarah explains, these "regenerative" relationships must be grounded in diversity and inclusivity:

> I'm comparing it to kind of the success or lack thereof in sustaining the health of a plot of land that's a complete monoculture. Like it can last for so long and be in good health for so long, but it's weak. [. . .] If you have this lovely collection of like-minded people that are all coming from the same kind of experience, the same perspective, that can be fun; like hang out with those people, work with those people, but it's so confined to that [. . .] one kind of environment, or perspective. [. . .] It can also be really detrimental in that it can create so much of this, like, elitist or exclusive environment that people that want to be involved may not feel very supported in being involved, and they won't be involved.—Sarah

Sarah's words parallel the thesis of the climate justice movement—that a broad-based social movement that inspires and supports diverse participants is our best hope for building a healthy and regenerative world. I hope that this book illuminates this thesis, and that by detailing how people work across lines in their daily lives to resist extreme energy extraction, it helps us realize climate justice.

Notes

1. See Clare Foran (2016) for information on Trump's climate-change denial and Kirk, Philbrick, and Roth (2016) for a compilation of Trump's derogatory statements.

2. See Popovich, Albeck-Ripka, and Pierre-Louise (2021) for an overview of Trump's rollback of environmental regulations, estimated to significantly increase greenhouse gas emissions (Pitt, Larsen, and Young 2020) and extra deaths from poor air quality (The State Energy and Environmental Impact Center 2019).

INTRODUCTION

1. Jim and Alma chose to use their real names in this research. I explain the rationale for this later in this chapter.

2. At minimum, I define *activist* as someone who spends time learning and sharing information about something they perceive as either unjust or just. See chapter 3 for more discussion of this term.

3. In her 2020 book, *Harvest the Vote*, Keystone XL pipeline resister Jane Kleeb affirms my thesis, arguing that it is necessary and possible for progressive movements to work with rural communities to address core challenges in the United States.

4. Collaboration, care, and interdependence are prioritized by women activists in diverse settings (Bell 2013; Brown and Ferguson 1995; Naples 1998; Taylor and Whittier 1992).

5. In recent years, Californians have been moving to Idaho in great numbers. In 2015, a quarter of all new residents moving to the state came from California (Siegler 2017). According to Siegler's (2017) NPR report, many are driven by a desire to move to more conservative communities. This characterizes the motivations to move of some of the core southwest Idaho anti-fracking interviewees who are California transplants. This spatial trajectory is also part of my own history. My parents are both from California.

6. Klare (2012) defines extreme energy extraction as any method requiring companies "to drill in extreme temperatures or extreme weather, or use extreme pressures, or operate under extreme danger—or some combination of these."

CHAPTER I. THE ENERGY AND POLITICAL LANDSCAPE

1. Bill McKibben (2016b), Ezra Silk (2016), and The Climate Mobilization (http://www.theclimatemobilization.org/) have argued that the United States should be mobilizing resources on the scale of World War II to address climate change.

2. See http://www.ucsusa.org/sites/default/files/images/2016/02/vehicles-fuels -extracting-tar-sands.jpg for a depiction of the difference between surface and in-situ tar sands mining.

3. See http://news.nationalgeographic.com/news/energy/2011/12/pictures /111222-canada-oil-sands-satellite-images/#/alberta-tar-oil-sands-satellite -pictures-2011_46161_600x450.jpg for images of tar sands tailings ponds.

4. See Saxifrage 2013 for a list of countries, most of which are located in the Global South.

5. See US Energy Administration (2017) for a map of these regions.

6. See https://www.youtube.com/watch?v=4FDuqYld8C8 for a good overview of the water protectors' efforts.

7. While Shell and ConocoPhillips relinquished Arctic drilling leases in 2016, due to "the unpredictable federal regulatory environment" (Shell quoted in Neuhauser 2016) and, likely, intense protest by kayaktivists in 2015 (see Brait 2015), the Trump administration pursued Arctic drilling as part of its America First Energy Plan (Gramer 2017).

8. In 2002, a coalition of organizations drafted the Bali Principles of Climate Justice in preparation for the World Summit on Sustainable Development held in Johannesburg, South Africa.

9. See Bond (2014) and Gray et al. (2021) for excellent discussions of the concept and movement.

10. Stacia Sydoriak added not *under* my back yard (NUMBY) to these perspectives (personal communication with Peter Hall).

11. Transitions from NIMBY to NIABY and the importance of NIMBYism as a starting point for activism are well documented (see Awâsis 2014; Beamish and Luebbers 2009; Boudet 2011; D'Arcy 2014; Devine-Wright 2009; Vandehey 2013).

12. The Coastal Gas Link pipeline and Line 3 both pushed construction through during the COVID-19 pandemic, while Indigenous peoples across Turtle Island faced disproportionate harm from the pandemic.

CHAPTER 2. THE ORGANIZING LANDSCAPE

1. California's 2014 population was 38.8 million whereas Idaho's was 1.634 million.

2. Historically, Texas was number one, followed by Alaska, California, and Louisiana. In 2014, North Dakota replaced Alaska as number two. In 2017, California fell to number four and from 2018 to 2020, it has been number seven behind Texas, North Dakota, New Mexico, Oklahoma, Colorado, and Alaska.

3. According to a report commissioned by the American Petroleum Institute, in 2010, women accounted for 19 percent of total employment in the combined oil and gas and petrochemical industries. Out of the 1.3 million new direct job opportunities that the report estimates for the 2010 to 2030 period, women "could account for 185 thousand," or 14 percent (IHS Global Inc. 2014:2).

4. After the election, the oil company Citadel Exploration sued San Benito County for 1.2 billion dollars. Citadel dropped the suit in March 2015.

5. Local chapters of 350.org independently choose their campaigns and organizing styles.

6. Successes at many levels since then (e.g., Monterey County and New York State, which banned fracking in December 2014), demonstrate that this fear was unfounded. See chapter 7 for details on successes that occurred in Santa Barbara County after 2014.

7. See http://sbcvote.com/Elect/Resources/Results11_2014/results-1.htm for election results.

8. In 2019, two out of four Idaho members of the US Congress were still climate deniers (Hardin and Moser 2019).

9. This is almost equivalent to the population of Santa Barbara County and its two neighboring counties, San Luis Obispo County and Ventura County, whose combined population was 1.57 million in 2019.

10. In response to a request for information on the threats oil and gas drilling poses to mortgages, a 2016 letter from Idaho's deputy attorney general (Guyon 2016) stated that a common provision in mortgages "prohibits the borrower from transferring any portion of the mortgaged property without the prior approval of the lender." Leasing the rights to a property's subsurface minerals without the consent of the lender violates this provision. In addition, "[m]ost residential mortgages also prohibit borrowers from (1) using or storing hazardous materials on the property; or (2) engaging in activities that (a) violate environmental laws; (b) may require an environmental cleanup; or (c) decrease the property's values because of the presence or release of hazardous substances. Extracting oil and gas from a property may violate one or both of these prohibitions." See also an interactive set of documents on mortgages and oil and gas leases compiled by the *New York Times*: http://www.nytimes.com /interactive/us/drilling-down-documents-8.html, and CAIA's website: https:// integrityandaccountability.org/resources/property-impacts/.

11. Title companies in Idaho are not required to notify buyers about mineral rights. Thus, many residents do not know if they own their mineral rights.

12. This begs the question, emergency for whom? The government-declared "emergency" is clearly for the oil and gas companies. It parallels the state

of emergency issued by North Dakota's governor in September 2016, which allowed militarized opposition to the water protectors at Standing Rock. The ACLU called the North Dakota governor's declaration "a state of emergency for civil rights" (see Cook 2016).

13. To get a sense of Highway 12, the size of the megaloads, and people involved, see: http://www.thisamericanland.org/news/reports-from-the-edge -megaloads-blocked-from-wild-scenic-river-route#.WIJE5juCuJU.

14. Some megaloads deviated from these routes, being transported to tar sands processing refineries in Montana or traveling by rail (not depicted on map 2), rather than road.

15. See chapter 7 for the full story of the megaloads.

CHAPTER 3. IDAHO PART 1

1. In *Our Roots Run Deep as Ironweed* (2013), one of Shannon Bell's interviewees, Maria Gunnoe, who is an anti-mountaintop-removal activist, describes how the coal companies took away her daughter's ability to feel safe in their home, which a coal company–induced mudslide had almost washed away. The company caused psychological trauma. Though not as extreme, ruining people's sense of trust is a similar detrimental effect of oil and gas.

2. In Santa Barbara, the oil and gas industry ripped Measure P apart for lack of clarity and then spent over six million dollars spreading false information. See chapter 7.

3. The Idaho Conservation League does much good work. In a state with few environmental organizations, I am thankful for ICL. My critique here is limited to their stance on natural gas regulations. In 2017, I was happy to see them bring attention to the issue with a post titled "Is Fracking in Idaho's Future?" (Hopkins 2017).

CHAPTER 4. IDAHO PART 2

1. Forced pooling is a method whereby an oil and gas company can ask a state to force a certain percentage of mineral rights owners in a certain area or "pool" to lease their mineral rights, against their will, if a certain percentage of other mineral rights owners have agreed. In Idaho, this threshold for integration is only 55 percent. This means owners of 45 percent of minerals have no choice over whether they want extraction to occur beneath their homes. People who do not own minerals have no say at any point.

2. In some ways, this parallels the environmental justice movement's efforts to redefine what environment is. Rather than nature or the outside world, as the conservation and mainstream environmental movement had long understood the term, environmental justice activists redefined *environment* to mean where people live, work, and play, bringing in concerns about people's health and livelihoods.

3. Stacia Sydoriak added not *under* my back yard (NUMBY) (personal communication with Peter Hall).

4. In 2019, California bill AB345 proposed setbacks of 2,500 feet to reduce public health risks (Ferrar 2019).

5. CAIA members like Jim and Alma see agribusiness, and Monsanto in particular, as on a par with the oil and gas industry in terms of violations of principles of integrity and accountability. They have built CAIA's organizational foundation to work for integrity and accountability in any context and see oil and gas pollution as one of *many* potential campaigns.

6. The web thickens. Just before he left the governorship to be President Bush's secretary of the interior in 2006, then Idaho governor Dirk Kempthorne facilitated J. R. Simplot's donation of his mansion to the state of Idaho, to be used as the governor's mansion (Yardley 2010). Supported by the timber industry and Simplot in his position as secretary of the interior, Kempthorne supported a number of environmentally damaging policies. The Bush administration that Kempthorne served is infamous for making the fracking boom possible through such policies as the Halliburton loophole, passed in 2005, which exempts oil and gas from environmental regulations like the Safe Drinking Water Act.

7. While 93 percent of Idahoans identify as "white alone," only 81.6 percent of Idahoans identify as "white alone, not Hispanic or Latino," suggesting that a good portion of Latinx individuals in Idaho identify as white. The numbers are similar in places where CAIA organizes.

8. In *Wastelanding: Legacies of Uranium Mining in Navajo Country* (2015), Traci Brynne Voyles asks the question, what is "wasteland" and who gets to decide? Places designated as wastelands by governments and industry are typically home to marginalized communities. They are not wastelands at all. Idaho is a case of locating industry—wastelanding—in white areas. In this way, fracking extends the discourse of wastelanding to new arenas.

9. For example, 350.org and Attac France (2015) write, "At every stage powerful forces—fossil fuel corporations, agro-business companies, financial institutions, dogmatic economists, skeptics and deniers, and governments in the thrall of these interests—stand in the way or promote false solutions." To the extent that they oppose climate justice legislation, climate skeptics do stand in the way of climate justice. However, their uncertainty about climate change, and even opposition to legislation, does not preclude them from *also* supporting activities that alleviate climate change. I do not mean to excuse climate skepticism, but to highlight how, relative to structures of climate change denial (see chapter 1 for my account of the campaign to obscure climate science, still endorsed by powerful politicians in the United States), everyday people who are skeptical about climate change pose little threat and may even be allies in other ways.

CHAPTER 5. WORKING ACROSS INTERSECTIONAL LINES

1. This chapter incorporates and expands upon Grosse 2019.

2. See Reed and Foran (2002:338) note 2 for the lineage of the concept.

3. Intersectionality is mentioned twice in this seven-hundred-page handbook of social movement theory, in a chapter on feminism and the women's movement. In this chapter, Ferree and Mueller's (2007) treatment of intersectionality resonates with that of interviewees—that individuals and organizations are

shaped by multiple "relations of power and injustice" (578). They note that the intersectionality of social movements is "not always acknowledged theoretically" (578).

4. See chapter 6 for how dividing communities is part of the oil and gas industry's playbook.

5. Participatory democratic forms of organizing do have limitations for achieving inclusivity. Doerr (2007) demonstrates that various structures (e.g., for accessing funding and deciding what topics are discussed) within the European Social Forum process exclude materially less privileged participants, particularly women from the Global South. Lack of translation excludes participants with fewer or no English skills (Doerr 2008; Grosse and Mark 2020). Polletta outlines how forms of organizing grounded in friendships can lead to exclusivity and resistance to formal mechanisms for equalizing power (2002:4).

CHAPTER 6. WORKING ACROSS ORGANIZATIONAL LINES

1. Rosewarne, Goodman, and Pearse (2014) argue that the climate crisis has upended notions of radicalism and pragmatism. In their view, pragmatism as usual "has become an impossible demand" because it leads to climate crisis (5).

2. For analysis of collaboration among grassroots activists who do have different racial, ethnic, and Indigenous identities, see chapter 8.

3. C4 refers to the nonprofit tax designation 501(c)(4). This is a designation for nonprofits that can engage in political and legislative activities. Its downside is that donations made to a 501(c)(4)s are not tax deductible for the donor. A 501(c)(3) organization, in contrast, can receive tax deductible donations, but can only engage in charitable work—activities like education and defending civil rights. Being able to deduct donations from federal taxes is attractive for large donors.

4. The Community Environmental Council in Santa Barbara is an exception to this, seeking "to move the Santa Barbara region away from dependence on fossil fuels in one generation."

5. "You don't fight fascism because you're going to win. You fight fascism because it is fascist" is an anarchist meme by Jean-Paul Sartre (1992).

6. bell hooks's book *Feminist Theory: From Margin to Center* (1984) inspired this section's heading.

7. The International Agency for Research on Cancer (2015) declared 2,4-D a possible human carcinogen in 2015. It is a known endocrine disrupter. See the National Resource Defense Council's article "2,4-D: The Most Dangerous Pesticide You've Never Heard of" (Sedbrook 2016) for a detailed account of the pesticide.

CHAPTER 7. TWO TALES OF STRUGGLE

1. In the case of the megaloads, 2014 marked the end of direct action, but not legal battles.

2. This section incorporates and expands upon Grosse 2017b.

3. In one instance reported by the *Missoulian*, a megaload held up traffic for fifty-nine minutes, just one of five blockages that were over twenty-nine minutes. In total, the megaload held up traffic for more than fifteen minutes (the maximum delay allowed) ten times that night (Briggeman 2011).

4. This section incorporates and expands upon Grosse 2017a.

5. For an example, see this "No on P" ad: https://www.youtube.com/watch?v=D45TdAk3Jnc.

6. The Workforce Investment Board of Santa Barbara (2012), which is staffed by the County of Santa Barbara, issued a 2012 report examining the top industry clusters in Santa Barbara County. The report identified agriculture, tourism, and wineries as the dominant industry cluster, accounting for 36,088 jobs, or 15 percent of the County's workforce. Only 566 jobs were documented in the energy and environment industry cluster, or 0.2 percent of the County workforce.

CHAPTER 8. LESSONS FROM MEASURE P AND THE MEGALOADS

1. Media regularly hailed California governor Jerry Brown as a climate hero. See Alexander (2015) for an example that highlights some of California's best environmental policies. Despite this, Brown supported fracking throughout his time as governor.

2. A 2017 ranking of states that make it easy for corporations to procure clean energy ranks Idaho as forty-eighth (Retail Industry Leaders Association, Information Technology Industry Council, and Clean Edge 2017).

3. See Bhadra (2013) on disaster scripting.

4. As I argue in chapter 4, interviewees' reasons for resisting natural gas development are more complicated than just protecting their backyards. I show that they are both more than NIMBY activists and that NIMBYism is not always a negative form of activism. See chapter 1 for a discussion of NIMBYism. While NIMBYism has been associated with efforts by privileged white communities to keep toxic infrastructure out of their backyards, with little concern for the people-of-color communities who suffer from this infrastructure, scholars and activists are increasingly thinking about how NIMBYism can be a springboard for justice-oriented activism that seeks to protect everyone's backyard from toxic materials.

5. On her toxic tours, Alma showed me natural gas wells, where the gas lines went directly under a popular swimming spot in the river, how roads were deteriorating because of truck traffic related to natural gas development, and where, if you were to slide out on a corner while driving on icy winter roads, you could easily hit natural gas pipes sticking out of the ground. "Toxic tours" are a way for community groups to increase public awareness of environmental injustices. Communities for a Better Environment, based in California, gives these tours regularly: http://www.cbecal.org/get-involved/toxic-tours/.

6. Some of these students expressed connection to their home places—where they grew up and where they still spent holidays and summers with their families.

7. There is also the question of defining stakeholders. People who are not in groups are also stakeholders—everyone is a stakeholder in climate change. Yet individuals who are not part of a group are rarely invited to tables, both because of difficulties identifying who these folks might be and because of a lack of preexisting relationships (whether individual or group level) of trust.

8. The composition of Santa Barbara's environmental organizations in 2017 (see table 2) confirms Arlo's fears that organizations were doing little to improve Latinx outreach.

9. Relational organizing is facilitated by horizontal organizing structures, but it is not equivalent. Rather than referring to structure, it refers to the character of groups—emphasizing the importance of relationships for the well-being of individual activists and their groups (see chapter 5 for an expanded discussion on relational organizing).

10. In predominantly Latinx districts of South County, opposition to Measure P was less extreme than in North County, but still present. For example, Old Town Goleta voted 55.15 percent against Measure P. Opposition to Measure P ranged from 49.48 percent to 57.84 percent on the Eastside of Santa Barbara.

CONCLUSION

1. While women are proven leaders in climate change adaptation and mitigation, they will be hard pressed to adapt to the temperature increases that would result from worst-case scenario emissions trajectories.

Bibliography

Alexander, Kurtis. 2015. "Gov. Jerry Brown Marches California Climate Agenda to Paris." *San Francisco Chronicle*, November 27 (http://www.sf chronicle.com/bayarea/article/Gov-Jerry-Brown-marches-California-climate -6660918.php).

Alter, Charlotte. 2014. "Voter Turnout in Midterm Elections Hits 72-Year Low." *Time*, April 1, 2015 (http://time.com/3576090/midterm-elections-turnout -world-war-two/).

AmazonWatch.org. et al. 2015a. "Keep Fossil Fuels in the Ground: A Declaration for the Health of Mother Earth" (http://amazonwatch.org/assets/files /2015-keep-the-oil-in-the-ground-english.pdf).

American Lung Association in California. 2016. "Oil Lobby Spending in California: 2016 Update #3." Retrieved October 24, 2017 (http://www.lung.org /local-content/california/documents/oil-industry-lobbying-2016.pdf).

Appalachian Voices. 2013. "Mountaintop Removal 101." Retrieved April 18, 2017 (http://appvoices.org/end-mountaintop-removal/mtr101/).

Associated Press. 2012. "Controversial Oil/Gas Drilling Bill, HB 464, Signed into Law." *Spokesman-Review*, March 23 (http://www.spokesman.com /blogs/boise/2012/mar/23/controversial-oilgas-drilling-bill-hb-464-signed -law/).

———. 2014. "Oil Trains Numbers in North Idaho Likely to Rise." *Missoulian*, April 14, 2015 (http://missoulian.com/news/state-and-regional/oil -trains-numbers-in-north-idaho-likely-to-rise/article_6687b194-a889-11e3 -a01b-0019bb2963f4.html).

Association of American Railroads. 2014. "Moving Crude by Rail." July (https://www.energy.gov/sites/prod/files/2014/08/f18/chicago_qermeeting_gray_statement.pdf)

Atkins, David. 2014. "Lessons from the Fracking Ban Fights of 2014." *Washington Monthly*, November 16 (https://washingtonmonthly.com/2014/11/16/lessons-from-the-fracking-ban-fights-of-2014/).

Awâsis, Sâkihitowin. 2014. "Pipelines and Resistance across Turtle Island." Pp. 253–66 in *A Line in the Tar Sands: Struggles for Environmental Justice*, edited by Toban Black, Stephen D'Arcy, Tony Weis, and Joshua Kahn Russell. Oakland, CA: PM Press.

Ballotpedia. 2014. "States." Retrieved March 18, 2015 (http://ballotpedia.org/States#State_governments).

Banerjee, Neela. 2015. "Exxon's Oil Industry Peers Knew about Climate Dangers in the 1970s, Too." *Inside Climate News*, December 22 (https://insideclimatenews.org/news/22122015/exxon-mobil-oil-industry-peers-knew-about-climate-change-dangers-1970s-american-petroleum-institute-api-shell-chevron-texaco).

Barker, Rocky. 2016. "Department of Lands Sticks with 300 Foot Setback for Oil and Gas Plants." *Idaho Statesman*, July 7 (http://www.idahostatesman.com/news/local/news-columns-blogs/letters-from-the-west/article88199222.html).

Beamish, Thomas D., and Amy J. Luebbers. 2009. "Alliance Building across Social Movements: Bridging Difference in a Peace and Justice Coalition." *Social Problems* 56(4):647–76.

Beck, Ulrich. 1992. *Risk Society: Towards a New Modernity*. London: Sage Publications.

Bedall, Philip, and Christoph Görg. 2014. "Antagonistic Standpoints: The Climate Justice Coalition Viewed in Light of a Theory of Societal Relationships with Nature." Pp. 44–65 in *Routledge Handbook of the Climate Change Movement*, edited by Matthias Dietz, and Heiko Garrelts. New York: Routledge.

Bell, Shannon Elizabeth. 2013. *Our Roots Run Deep as Ironweed: Appalachian Women and the Fight for Environmental Justice*. Urbana: University of Illinois Press.

———. 2014. "'Sacrificed So Others Can Life Conveniently': Social Inequality, Environmental Injustice, and the Energy Sacrifice Zone of Central Appalachia." Pp. 261–74 in *Understanding Diversity: Celebrating Difference, Challenging Inequality*, edited by Claire M. Renzetti, and Raquel Kennedy Bergen. Boston, MA: Allyn and Bacon.

———. 2016. *Fighting King Coal: The Challenges to Micromoblization in Central Appalachia*. Cambridge, MA: The MIT Press.

Bell, Shannon Elizabeth, and Yvonne A. Braun. 2010. "Coal, Identity, and the Gendering of Environmental Justice Activism in Central Appalachia." *Gender & Society* 24(6):794–813.

Bergman, Noam. 2018. "Impacts of the Fossil Fuel Divestment Movement: Effects on Finance, Policy and Public Discourse." *Sustainability* 10(7):25–29.

Bevington, Douglas, and Chris Dixon. 2005. "Movement-Relevant Theory: Rethinking Social Movement Scholarship and Activism." *Social Movement Studies* 4(3):185–208.

Bhadra, Monamie. 2013. "Disaster Scripting in India's Nuclear Energy Landscape." STS Forum on the 2011 Fukushima / East Japan Disaster, May 11–14, Berkeley, CA (https://fukushimaforum.wordpress.com/workshops/sts-forum -on-the-2011-fukushima-east-japan-disaster/manuscripts/session-4a-when -disasters-end-part-i/disaster-scripting-in-indias-nuclear-energy-landscape/).

Bhavnani, Kum-Kum. 1993. "Tracing the Contours: Feminist Research and Feminist Objectivity." *Women's Studies International Forum* 16(2):95–104.

Bhavnani, Kum-Kum, and Krista Bywater. 2009. "Dancing on the Edge: Women, Culture, and a Passion for Change." Pp. 52–66 in *On the Edges of Development: Cultural Interventions*, edited by Kum-Kum Bhavnani, John Foran, Priya A. Kurian, and Debashish Munshi. New York: Routledge.

Bishop, Bill. 2008. *The Big Sort: Why the Clustering of Like-Minded America Is Tearing Us Apart*. Boston: Houghton Mifflin Company.

Black, Toban, Stephen D'Arcy, Tony Weis, and Joshua Kahn Russell, eds. 2014a. *A Line in the Tar Sands: Struggles for Environmental Justice*. Oakland, CA: PM Press

———. 2014b. "Introduction." Pp. 1–20 in *A Line in the Tar Sands: Struggles for Environmental Justice*, edited by Toban Black, Stephen D'Arcy, Tony Weis, and Joshua Kahn Russell. Oakland, CA: PM Press.

Blee, Kathy. 2012. *Democracy in the Making: How Activist Groups Form*. Oxford: Oxford University Press.

Boggs, Carl. 1977–78. "Marxism, Prefigurative Communism and the Problem of Workers' Control." *Radical America* 11–12(6, 1):99–122.

Bond, Patrick. 2014. "Justice." Pp. 133–45 in *Critical Environmental Politics: Interventions*, edited by Carl Death. London: Routledge.

Boudet, Hilary Schaffer 2011. "From NIMBY to NIABY: Regional Mobilization against Liquefied Natural Gas in the United States." *Environmental Politics* 20(6):786–806.

Bowles, Paul, and Henry Veltmeyer, eds. 2014. *The Answer Is Still No: Voices of Pipeline Resistance*. Halifax, NS: Fernwood Publishing.

Brait, Ellen. 2015. "Portland's Bridge-Hangers and 'Kayaktivists' Claim Win in Shell Protest." *The Guardian*, July 31 (https://www.theguardian.com /business/2015/jul/31/portland-bridge-shell-protest-kayaktivists-fennica -reaction).

Breines, Wini. 1982. *Community and Organization in the New Left: 1962– 1968: The Great Refusal*. New Brunswick, NJ: Rutgers University Press.

Briggeman, Kim. 2011. "Snowy Roads, Traffic Delay Violations Stall ConocoPhillips Megaloads." *Missoulian*, February 5 (http://missoulian.com/news /local/snowy-roads-traffic-delay-violations-stall-conocophillips-megaloads /article_b2f65056-307a-11e0-8332-001cc4c002e0.html).

———. 2015. "Kearl Oil Sands Expands Early; CEO Credits Lessons from Protests." *Missoulian*, June 21 (http://missoulian.com/news/local/kearl-oil -sands-expansion-ramps-up-early-ceo-credits-lessons/article_369c04ef-4904 -5234-baa8-64f5539f28b9.html).

Broder, John M. 2009. "Gore's Dual Role: Advocate and Investor." *New York Times*, November 2 (http://www.nytimes.com/2009/11/03/business/energy -environment/03gore.html).

Brown, Alleen. 2016. "Fracking Pipeline Puts Tim Kaine's Fossil Fuel Industry Ties to the Test." *EcoWatch*, August 3 (http://www.ecowatch.com/a-fracking-pipeline-puts-tim-kaines-fossil-fuel-industry-ties-to-the-t-19484 04158.html).

Brown, Phil, and Faith I. T. Ferguson. 1995. "'Making a Big Stink': Women's Work, Women's Relationships, and Toxic Waste Activism." *Gender & Society* 9(2):145–72.

Brown, Phil, and Edwin Mikkelsen. 1990. *No Safe Place: Toxic Waste, Leukemia, and Community Action*. Berkeley: University of California Press.

Brown, Stephen P. A., and Mine K. Yucel. 2013. "The Shale Gas and Tight Oil Boom: U.S. States' Economic Gains and Vulnerabilities." Retrieved August 7, 2021 (https://www.cfr.org/report/shale-gas-and-tight-oil-boom).

Bueckert, Kate. 2016. "Enbridge Line 9 Pipeline Appeal Approved by Supreme Court." CBC News, March 10 (http://www.cbc.ca/news/canada/kitchener-waterloo/line-9-enbridge-pipeline-appeal-approved-scoc-1.3485081).

Burgess, Robert 1984. "Methods of Field Research 2: Interviews as Conversations." Pp. 101–22 in *In the Field: An Introduction to Field Research*. London: Macmillan.

Bystydzienski, Jill M., and Steven P. Schacht, eds. 2001. *Forging Radical Alliances Across Difference: Coalition Politics for the New Millennium*. Lanham, MD: Rowman and Littlefield.

Cabello, Joanna, and Tamra Gilbertson. 2012. "A Colonial Mechanism to Enclose Lands: A Critical Review of Two REDD+-Focused Special Issues." *Ephemera* 12(1/2):162–80.

Cable, Sherry, and Charles Cable. 1995. *Environmental Problems/Grassroots Solutions: The Politics of Environmental Conflict*. New York: St. Martin's.

Cable, Sherry, Tamara L. Mix, and Donald W. Hastings. 2005. "Mission Impossible? Environmental Justice Activists' Collaboration with Professional Environmentalists and with Academics." Pp. 55–76 in *Power, Justice and the Environment: A Critical Appraisal of the Environmental Justice Movement*, edited by David N. Pellow, and Robert J. Brulle. Cambridge, MA: MIT Press.

CAIA. 2016. "Citizens Allied for Integrity and Accountability." Retrieved March 11, 2016 (http://integrityandaccountability.org/).

California Student Sustainability Coalition. 2014. "CSSC In Solidarity with Ferguson." Retrieved: October 8 2015 (http://www.sustainabilitycoalition.org/cssc-in-solidarity-with-ferguson/).

Campbell, James E. 1987. "The Revised Theory of Surge and Decline." *American Journal of Political Science* 31(4):965–79.

Caniglia, Beth Schaefer, Robert J. Brulle, and Andrew Szasz. 2015. "Civil Society, Social Movements, and Climate Change." Pp. 235–68 in *Climate Change and Society: Sociological Perspectives*, edited by Riley E. Dunlap, and Robert J. Brulle. New York, NY: Oxford University Press.

Cart, Julie. 2015a. "Oil Well Oversight in L.A. Basin Is 'Inconsistent,' Audit Finds." *Los Angeles Times*, October 8 (http://www.latimes.com/local/california/la-me-oil-report-health-20151009-story.html).

———. 2015b. "State Issues Toughest-in-the-Nation Fracking Rules." *Los Angeles Times*, July 1 (http://www.latimes.com/local/lanow/la-me-ln-state-issues-fracking-rules-20150701-story.html).

———. 2015c. "High Levels of Benzene Found in Fracking Waste Water." *Los Angeles Times*, February 11 (http://www.latimes.com/local/california/la-me-fracking-20150211-story.html#page=1).

Center for Responsive Politics. "JR Simplot Co." OpenSecrets. Retrieved March 12, 2016 (https://www.opensecrets.org/pacs/lookup2.php?strID=C00120873&cycle=2012).

Cho, Renee. 2011. "Mountaintop Removal: Laying Waste to Streams and Forests." *State of the Planet* (blog) (http://blogs.ei.columbia.edu/2011/08/11/mountaintop-removal-laying-waste-to-streams-and-forests/).

Chow, Lorraine. 2015. "It's Official: Oklahoma Experiences More Earthquakes Than Anywhere Else in the World." *EcoWatch*, November 16 (http://ecowatch.com/2015/11/16/oklahoma-most-earthquakes-fracking/).

Christian, Sena. 2016. "Bunker Hill Superfund Site Is Still a Toxic Mess, with Legacy of Suffering." *Newsweek*, June 12 (http://www.newsweek.com/2016/06/24/bunker-hill-superfund-silver-valley-lead-poisoning-469222.html).

CIRCLE (Center for Information and Research on Civic Learning and Engagement). 2015. "2014 Youth Turnout and Youth Registration Rates Lowest Ever Recorded; Changes Essential in 2016." Tufts University Jonathan M. Tisch College of Civic Life (http://civicyouth.org/2014-youth-turnout-and-youth-registration-rates-lowest-ever-recorded-changes-essential-in-2016/).

Clark, Brett, and Richard York. 2005. "Carbon Metabolism: Global Capitalism, Climate Change, and the Biospheric Rift." *Theory and Society* 34(4):391–428.

Committee on Energy and Commerce. 2011. *Chemicals Used in Hydraulic Fracturing*. United States House of Representatives (ecolo.org/documents/documents_in_english/gas-_Hydraulic-Fract-chemicals-2011-report.pdf).

Concerned Health Professionals of NY and Physicians for Social Responsibility. 2016. "Compendium of Scientific, Medical, and Media Findings Demonstrating Risks and Harms of Fracking (Unconventional Gas and Oil Extraction)" (http://concernedhealthny.org/compendium/).

Cook, Jennifer. 2016. "North Dakota's Governor Declared a State of Emergency to Deal with Peaceful Oil Pipeline Protesters. We Call It a State of Emergency for Civil Rights." ACLU (https://www.aclu.org/blog/speak-freely/north-dakotas-governor-declared-state-emergency-deal-peaceful-oil-pipeline).

County of Santa Barbara. 2013. "Job Opportunity for County of Santa Barbara." Retrieved March 15, 2017 (http://cosb.countyofsb.org/uploadedFiles/hr/Employment/CSB_Brochure.pdf).

County of Santa Barbara Staff. 2014. *Impact Analysis Report on Initiative to Ban "High-Intensity Petroleum Operations."* Responding to Elections Code 9111: Attachment A.

Crenshaw, Kimberlé. 1989. "Demarginalizing the Intersection of Race and Sex: A Black Feminist Critique of Antidiscrimination Doctrine, Feminist Theory and Antiracist Politics." *University of Chicago Legal Forum* 140:139–67.

Curtis, Morgan. 2015. "19 Ways You Can Work for Climate Justice." SustainUS (http://sustainus.org/2015/11/19-ways-you-can-work-for-climate-justice/).

Dal Bó, Ernesto. 2006. "Regulatory Capture: A Review." *Oxford Review of Economic Policy* 22(2):203–25.

Dalrymple, Amy. 2016. "Pipeline Route Plan First Called for Crossing North of Bismarck." *Bismark Tribune*, August 18 (http://bismarcktribune.com /news/state-and-regional/pipeline-route-plan-first-called-for-crossing-north -of-bismarck/article_64d053e4-8a1a-5198-a1dd-498d386c933c.html).

Dankelman, Irene. 2010. *Gender and Climate Change: An Introduction*. New York: Earthscan.

D'Arcy, Stephen. 2014. "Secondary Targetting: A Strategic Approach to Tar Sands Resistance." Pp. 286–96 in *A Line in the Tar Sands: Struggles for Environmental Justice*, edited by Toban Black, Stephen D'Arcy, Tony Weis, and Joshua Kahn Russell. Oakland, CA: PM Press.

Davis, Lynne, ed. 2010. *Alliances: Re/Envisioning Indigenous–Non-Indigenous Relationships*. Toronto: University of Toronto Press.

della Porta, Donatella, and Louisa Parks. 2014. "Framing Processes in the Climate Movement: From Climate Change to Climate Justice." Pp. 19–30 in *Routledge Handbook of the Climate Change Movement*, edited by Matthias Dietz, and Heiko Garrelts. New York: Routledge.

De Lucia, Vito. 2014. "The Climate Justice Movement and the Hegemonic Discourse of Technology." Pp. 66–83 in *Routledge Handbook of the Climate Change Movement*, edited by Matthias Dietz, and Heiko Garrelts. New York: Routledge.

Democracy Now! 2016. "Standing Rock: 100+ Injured after Police Attack with Water Cannons, Rubber Bullets and Mace." *Democracy Now!*, November 21 (https://www.democracynow.org/2016/11/21/headlines/standing_rock_100 _injured_after_police_attack_with_water_cannons_rubber_bullets_mace).

Devine-Wright, Patrick. 2009. "Rethinking NIMBYism: The Role of Place Attachment and Place Identity in Explaining Place-Protective Action." *Journal of Community and Applied Social Psychology* 19(6):426–41.

Division of Oil, Gas and Geothermal Resources. 2013a. "DOGGR: Drilling Through Time." California Department of Conservation. Retrieved March 17, 2015 (http://www.conservation.ca.gov/index/aboutus/Pages/aboutus_doggr .aspx).

———. 2013b. "Well Counts and Production of Oil, Gas and Water By County 2013." California Department of Conservation.

Doerr, Nicole. 2007. "Is 'Another' Public Sphere Actually Possible? The Case of 'Women Without' in the European Social Forum Process as a Critical Test for Deliberative Democracy." *Journal of International Women's Studies* 8(3):71–88.

———. 2008. "Deliberative Discussion, Language and Efficiency in the World Social Forum Process." *Mobilization: An International Journal* 13(4):395–410.

Drake, Bruce. 2015. "How Americans View the Top Energy and Environmental Issues." The Pew Research Center (http://www.pewresearch.org/fact -tank/2015/01/15/environment-energy-2/).

Dunlap, Riley E., and Robert J. Brulle, eds. 2015. *Climate Change and Society: Sociological Perspectives*. New York: Oxford University Press.

Dunlap, Riley E., and Peter J. Jacques. 2013. "Climate Change Denial Books and Conservative Think Tanks." *American Behavioral Scientist* 57(6):699–731.

Ehrlich, April. 2014. "'Fractivist' Enters Sixth Day of Hunger Strike." *Independent-Enterprise*, October 15 (http://www.argusobserver.com /independent/news/fractivist-enters-sixth-day-of-hunger-strike/article _acd3972e-5498-11e4-a4dd-db0985e6e4ae.html).

Eligon, John. 2013. "An Oil Town Where Men Are Many, and Women are Hounded." *New York Times*, January 15 (http://www.nytimes.com/2013 /01/16/us/16women.html?pagewanted=all).

Ellingboe, Kristen, and Ryan Koronowski. 2016. "Most Americans Disagree with Their Congressional Representative on Climate Change." *Think Progress* (https://archive.thinkprogress.org/most-americans-disagree-with-their -congressional-representative-on-climate-change-95dcoeee7b8f/).

Escobar, Arturo. 1992. "Imagining a Post-Development Era? Critical Thought, Development and Social Movements." *Social Text* (31/32):20–56.

———. 2001. "Culture Sits in Places: Reflections on Globalism and Subaltern Strategies of Localization." *Political Geography* 20(2):139–74.

Estes, Nick. 2019. *Our History Is the Future: Standing Rock versus the Dakota Access Pipeline, and the Long Tradition of Indigenous Resistance*. New York: Verso.

Estes, Nick, and Jaskiran Dhillon, eds. 2019. *Standing with Standing Rock: Voices from the #NoDAPL Movement*. Minneapolis: University of Minnesota Press.

Feld, Steven, and Keith H. Basso. 1996. *Senses of Place*. Santa Fe, NM: School of American Research Press.

Ferrar, Kyle. 2014. "Hydraulic Fracturing Stimulations and Oil Drilling Near California Schools and within School Districts Disproportionately Burdens Hispanic and Non-white Students." FracTracker Alliance (https:// www.fractracker.org/a5ej2osjfwe/wp-content/uploads/2014/11/Fractracker _SchoolEnrollmentReport_11.17.14.pdf).

———. 2019. "Impact of a 2,500' Oil and Gas Well Setback in California." FracTracker Alliance. Retrieved: April 23, 2020 (https://www .fractracker.org/2019/07/impact-of-a-2500-oil-and-gas-well-setback-in -california/).

Ferree, Myra Marx, and Carol McClurg Mueller. 2007. "Feminism and the Women's Movement: A Global Perspective." Pp. 576–607 in *The Blackwell Companion to Social Movements*, edited by David A. Snow, Sarah A. Soule, and Hanspeter Kriesi. Malden, MA: Blackwell Publishing.

Food and Water Watch. 2013. *The Social Costs of Fracking: A Pennsylvania Case Study*. Washington, DC: Food and Water Watch.

Foran, Clare. 2016. "Donald Trump and the Triumph of Climate-Change Denial." *The Atlantic*, December 25 (https://www.theatlantic.com/politics /archive/2016/12/donald-trump-climate-change-skeptic-denial/510359/).

Foran, John. 2014. "Beyond Insurgency to Radical Social Change: The New Situation." *Social Justice Studies* 8(1):5–25.

———. "The Paris Agreement: Paper Heroes Widen the Climate Justice Gap." Resilience (https://www.resilience.org/stories/2015-12-14/the-paris-agreement-paper-heroes-widen-the-climate-justice-gap/).

———. 2016. "A Few Thoughts on Studying the Most Radical Social Movement of the Twenty-First Century." Resilience (http://www.resilience.org/stories/2016-03-14/a-few-thoughts-on-studying-the-most-radical-social-movement-of-the-twenty-first-century).

Foran, John, and Richard Widick. 2013. "Breaking Barriers to Climate Justice." *Contexts* 12(2):34–49.

Forbes. 2016a. "World's Biggest 25 Oil and Gas Companies in 2016." May 26 (https://www.forbes.com/pictures/flhm45edljk/schlumberger/#2972a03a1467).

———. 2016b. "The World's Biggest Public Energy Companies 2016." May 30 (https://www.forbes.com/pictures/hefj45fim/no-3-exxonmobil/#4bfdoaf26e4d).

Foster, John Bellamy, Brett Clark, and Richard York. 2010. *The Ecological Rift: Capitalism's War on the Earth*. New York: Monthly Review Press.

Fox, Julia. 1999. "Mountaintop Removal in West Virginia: An Environmental Sacrifice Zone." *Organization & Environment* 12(2):163–83.

Foytlin, Cherri, Yudith Nieto, Kerry Lemon, and Will Wooten. 2014. "Gulf Coast Resistance and the Southern Leg of the Keystone XL Pipeline." Pp. 181–94 in *A Line in the Tar Sands: Struggles for Environmental Justice*, edited by Toban Black, Stephen D'Arcy, Tony Weis, and Joshua Kahn Russell. Oakland, CA: PM Press.

FracTracker Alliance. 2017. "Oil and Gas Drilling 101" (https://www.fractracker.org/resources/oil-and-gas-101/).

Friends of the Clearwater. N.d. "Important US 12 Megaload Update." Retrieved August 7, 2021 (https://www.friendsoftheclearwater.org/newsletters/important-us-12-megaload-update/).

Geertz, Clifford. 1973. *The Interpretation of Cultures: Selected Essays*. New York: Basic Books, Inc., Publishers.

———. 1996. "Afterward." Pp. 259–62 in *Senses of Place*, edited by Steven Feld, and Keith H. Basso. Santa Fe, NM: School of American Research Press.

Gelman, Andrew. 2011. "Why America Isn't as Polarized as You Think." *The Atlantic*, July 8 (https://www.theatlantic.com/politics/archive/2011/07/why-america-isnt-as-polarized-as-you-think/240653/).

Geredien, Ross. 2009. "Post-Mountaintop Removal Reclamation of Mountain Summits for Economic Development in Appalachia." Natural Resources Defense Council (http://www.ilovemountains.org/reclamation-fail/mining-reclamation-2010/MTR_Economic_Reclamation_Report_for_NRDC_V7.pdf).

Gilbert, Daniel. 2014. "Exxon CEO Joins Suit Citing Fracking Concerns." *Wall Street Journal*, February 20 (http://www.wsj.com/articles/SB10001424052702304899704579391181466603804).

Gilio-Whitaker, Dina. 2019. *As Long as Grass Grows: The Indigenous Fight for Environmental Justice: From Colonization to Standing Rock*. Boston: Beacon.

Gillis, Justin, and John Schwartz. 2015. "Deeper Ties to Corporate Cash for Doubtful Climate Researcher." *New York Times*, February 21 (https://www.nytimes.com/2015/02/22/us/ties-to-corporate-cash-for-climate-change-researcher-Wei-Hock-Soon.html).

Goffman, Erving. 1974. *Frame Analysis: An Essay on the Organization of Experience.* Cambridge, MA: Harvard University Press.

Goodman, Amy. 2016. "Standing Rock Special: Unlicensed #DAPL Guards Attacked Water Protectors with Dogs and Pepper Spray." *Democracy Now!*, November 24 (https://www.democracynow.org/2016/11/24/standing_rock_special_unlicensed_dapl_guards).

Gramer, Robbie. 2017. "Oil Companies Cool on Arctic Drilling. Trump Wants It Anyway." *Foreign Policy*, March 24 (http://foreignpolicy.com/2017/03/24/oil-companies-cool-on-arctic-drilling-trump-wants-it-anyway-energy-alaska-environment/).

Gray, Summer, Corrie Grosse, Brigid Mark, and Erica Morrell. 2021. "Climate Justice Movements and Sustainable Development Goals." Pp. 1–10 in *Climate Action*, edited by Walter Leal Filho, Anabela Marisa Azul, Luciana Brandli, Pinar Gökcin Özuyar, and Tony Wall. New York: Springer International Publishing.

Grosse, Corrie. 2017a. "Megaloads and Mobilization: The Rural People of Idaho Stand against Big Oil." *Case Studies in the Environment* 1(1):1–7. https://doi.org/10.1525/cse.2017.sc.450285.

———. 2017b. "Grassroots vs. Big Oil: Measure P and the Fight to Ban Fracking in Santa Barbara County, California." *Case Studies in the Environment* 1(1):1–6. https://doi.org/10.1525/cse.2017.sc.442387.

———. 2019. "Climate Justice Movement Building: Values and Cultures of Creation in Santa Barbara, California." *Social Sciences* 8(79):1–26.

Grosse, Corrie, and Brigid Mark. 2020. "A Colonized COP: Indigenous Exclusion and Youth Climate Justice Activism at the United Nations Climate Change Negotiations." *Journal of Human Rights and the Environment* (11):146–70.

Grossman, Zoltán. 2017. *Unlikely Alliances: Native Nations and White Communities Join to Defend Rural Lands.* Seattle: University of Washington Press.

Guyon, Stephanie. 2016. "Impact of Oil and Gas Leases on Mortgages—Title 47, Chapter 3, Idaho Code." Edited by Michelle Stennett, Office of the Attorney General.

Han, Hari. 2014. *How Organizations Develop Activists.* New York: Oxford University Press.

Hansen, James. 2012. "Game Over for the Climate." *New York Times*, May 9 (http://www.nytimes.com/2012/05/10/opinion/game-over-for-the-climate.html).

Haraway, Donna. 1988. "Situated Knowledges: The Science Question in Feminism and the Privilege of Partial Perspective." *Feminist Studies* 14(3):575–99.

Hardin, Sally, and Claire Moser. 2019. "Climate Deniers in the 116th Congress." *Center for American Progress Action Fund.* Retrieved April 23, 2020 (https://www.americanprogressaction.org/issues/green/news/2019/01/28/172944/climate-deniers-116th-congress/).

Harlan, Sharon L., David N. Pellow, J. Timmons Roberts, Shannon Elizabeth Bell, William G. Holt, and Joane Nagel. 2015. "Climate Justice and Inequality." Pp. 127–63 in *Climate Change and Society: Sociological Perspectives*, edited by Riley E. Dunlap, and Robert J. Brulle. New York: Oxford University Press.

Hatch, Christopher, and Matt Price. 2008. "Canada's Toxic Tar Sands: The Most Destructive Project on Earth." Toronto: Environmental Defence (http://environmentaldefence.ca/wp-content/uploads/2016/01/TarSands _TheReport.pdf).

Healy, Jack. 2015. "Heavyweight Response to Local Fracking Bans." *New York Times*, January 3 (https://www.nytimes.com/2015/01/04/us/heavyweight -response-to-local-fracking-bans.html).

Herman, Edward S., and Noam Chomsky. [1988] 2002. *Manufacturing Consent: The Political Economy of the Mass Media*. New York: Pantheon Books.

Herr, Alexandria. 2021. "One of the Most Polluted Counties in America Is Getting 40,000 More Oil Wells." Grist, March 11 (https://grist.org/justice /one-of-the-most-polluted-counties-in-america-is-getting-40000-more-oil -wells/).

Hoggan, James. 2016. *I'm Right and You're an Idiot: The Toxic State of Public Discourse and How to Clean It Up*. Gabriola Island, BC, Canada: New Society Publishers.

Holdren, John P. 2014. "The Overwhelming Consensus of Climate Scientists Worldwide." The White House. Retrieved February 5, 2017 (https:// obamawhitehouse.archives.gov/blog/2014/06/20/overwhelming-consensus -climate-scientists-worldwide).

Honor the Earth. N.d. "Man Camps Fact Sheet" (http://www.honorearth.org /man_camps_fact_sheet).

hooks, bell. 1984. *Feminist Theory: From Margin to Center*. Boston: South End Press.

Hopkins, Austin. 2017. "Is Fracking in Idaho's Future?" *Idaho Conservation League Blog* (http://www.idahoconservation.org/blog/fracking-idahos -future/).

Hormel, Leontina M. 2016. "Nez Perce Defending Treaty Lands in Northern Idaho." *Peace Review* 28(1):76–83.

Horn, Paul. 2016. "Map: The Fracking Boom, State by State." *InsideClimate News* (https://insideclimatenews.org/content/map-fracking-boom-state-state).

Howarth, Robert W., Renee Santoro, and Anthony Ingraffea. 2011. "Methane and the Greenhouse-Gas Footprint of Natural Gas from Shale Formations." *Climatic Change* 106(4):679.

Idaho Conservation League. 2015. "Annual Report 2015." Retrieved March 9, 2016 (https://www.idahoconservation.org/wp-content/uploads/2020/07/ICL 2015annrpt.pdf).

Idaho Department of Lands. 2015. "Mineral Rights Integration Information Sheet." Retrieved February 4, 2015 (https://www.landcan.org/pdfs/2015-01 _faq-mineral-rights-integration_v0204%20(1).pdf).

———. 2017. "Oil and Gas Drill Permits." Retrieved February 15, 2017 (https://www.idl.idaho.gov/oil-gas/regulatory/well-permits/index.html).

Idaho Oil and Gas Conservation Commission. 2016. "Strategic Plan 2017–2018."

Idaho Petroleum Council. 2016. "Letter of Support for S1339."

IHS Global Inc. 2014. "Minority and Female Employment in the Oil and Gas and Petrochemical Industries." Washington, DC: IHS Global Inc. (http://www.api.org/~/media/files/policy/jobs/ihs-minority-and-female-employment -report.pdf).

InsideClimate News. 2015. "Exxon: The Road Not Taken" (https://inside climatenews.org/content/Exxon-The-Road-Not-Taken).

International Agency for Research on Cancer. 2015. "IARC Monographs Evaluate DDT, Lindane, and 2,4-D." Press Release 236. World Health Organization, June 23 (http://www.iarc.fr/en/media-centre/pr/2015/pdfs/pr236_E.pdf).

International Energy Agency. 2015. "Energy and Climate Change: World Energy Outlook Special Briefing for COP21" (http://climateknowledge.org /figures/Rood_Climate_Change_AOSS480_Documents/IEA_COP21_Paris _Briefing_IntEnerAgen_2015.pdf)

IPCC (Intergovernmental Panel on Climate Change). 2014. "Summary for Policymakers." In *Climate Change 2014: Impacts, Adaptation, and Vulnerability*, edited by C. B. Field, V. R. Barros, D. J. Dokken, K. J. Mach, M. D. Mastrandrea, T. E. Bilir, M. Chatterjee, K. L. Ebi, Y. O. Estrada, R. C. Genova, B. Girma, E. S. Kissel, A. N. Levy, S. MacCracken, P. R. Mastrandrea, and L. L. White. Cambridge: Cambridge University Press (https://www.ipcc.ch/report/ar5/wg2/).

———. 2018. "Summary for Policymakers." In *Global Warming of 1.5°C*, edited by V. Masson-Delmotte, P. Zhai, H. O. Pörtner, D. Roberts, J. Skea, P. R. Shukla, A. Pirani, W. Moufouma-Okia, C. Péan, R. Pidcock, S. Connors, J. B. R. Matthews, Y. Chen, X. Zhou, M. I. Gomis, E. Lonnoy, T. Maycock, M. Tignor, and T. Waterfield. Geneva, Switzerland: World Meteorological Organization (https://www.ipcc.ch/sr15/).

———. 2021. "Summary for Policy Makers." In *Climate Change 2021: The Physical Science Basis*, edited by V. Masson-Delmotte, P. Zhai, A. Pirani, S. L. Connors, C. Péan, S. Berger, N. Caud, Y. Chen, L. Goldfarb, M. I. Gomis, M. Huang, K. Leitzell, E. Lonnoy, T. K. Maycock, T. Waterfield, O. Yelekçi, R. Yu, and B. Zhou. Cambridge: Cambridge University Press (https://www .ipcc.ch/report/sixth-assessment-report-working-group-i/).

Jennings, Katie, Dino Grandoni, and Susanne Rust. 2015. "How Exxon Went from Leader to Skeptic on Climate Change Research." *Los Angeles Times*, October 23 (http://graphics.latimes.com/exxon-research/).

Jerving, Sara, Katie Jennings, Masako Melissa Hirsch, and Susanne Rust. 2015. "What Exxon Knew about the Earth's Melting Arctic." *Los Angeles Times*, October 9 (http://graphics.latimes.com/exxon-arctic/).

Johnson, Jean Kirsten. 2018. "Nez Perce Commemorate Confrontation's Anniversary." *Indian Country Today*, September 12 (https://indiancountrytoday .com/archive/nez-perce-commemorate-confrontations-anniversary)

Johnson, Kirk. 2013. "Fight over Energy Finds a New Front in a Corner of Idaho." *New York Times*, Sept 25 (http://www.nytimes.com/2013/09/26/us /fight-over-energy-finds-a-new-front-in-a-corner-of-idaho.html).

Juris, Jeffrey S., Erica G. Bushell, Meghan Doran, J. Matthew Judge, Amy Lubitow, Bryan Maccormack, and Christopher Prener. 2014. "Movement Building and the United States Social Forum." *Social Movement Studies* 13(3):328–48.

Kelley, Robin D. G. 2002. *Freedom Dreams: The Black Radical Imagination.* Boston: Beacon Press.

Kelso, Matt. 2015. "1.7 Million Wells in the U.S.—A 2015 Update" FracTracker Alliance, August 3 (https://www.fractracker.org/2015/08/1-7-million-wells/).

King, Martin Luther Jr. 1967. "Beyond Vietnam: A Time to Break Silence." Retrieved February 27 2019 (http://www.americanrhetoric.com/speeches /mlkatimetobreaksilence.htm).

Kirk, Chris, Ian Prasad Philbrick, and Gabriel Roth. 2016. "230 Things Donald Trump Has Said and Done That Make Him Unfit to Be President." *Slate*, November 7 (http://www.slate.com/articles/news_and_politics/cover_story /2016/07/donald_trump_is_unfit_to_be_president_here_are_141_reasons _why.html).

Klare, Michael T. 2012. "The New "Golden Age of Oil" That Wasn't." *Tom-Dispatch* (http://www.tomdispatch.com/post/175601/tomgram%3A_michael _klare,_extreme_energy_means_an_extreme_planet/).

Kleeb, Jane. 2020. *Harvest the Vote: How Democrats Can Win Again in Rural America.* New York: HarperCollins Publishers.

Klein, Naomi. 2014. *This Changes Everything: Capitalism vs. the Climate.* New York: Simon & Schuster.

Koch, Blair. 2014. "Oil Company VP/Ex-Halliburton Fracking Lobbyist Assaults EnviroNews Chief Outside Public Meeting." *EnviroNews Idaho*, November 7 (http://www.environews.tv/111714-alta-mesa-vp-halliburton-fracking -lobbyist-assaults-environews-chief-outside-public-meeting/).

———. 2015. "Fracking Activist Alma Hasse Files $1.5 Million Lawsuit for 'Wrongful Arrest' after All Charges Dropped." *EnviroNews Idaho*, April 8 (http://environews.tv/040815-fracking-activist-alma-hasse-files-1-5-million -lawsuit-for-wrongful-arrest-after-all-charges-dropped/).

Krauss, Celene. 1989. "Community Struggles and the Shaping of Democratic Consciousness." *Sociological Forum* 4:227–39.

Krosnick, John. 2013. "What Do the Residents of Various States Believe about Global Warming?" Stanford University. Retrieved February 17, 2017 (http:// climatepublicopinion.stanford.edu/sample-page/opinions-in-the-states/#maps).

Krueger, Alyson. 2019. "'Unlikely' Hikers Hit the Trail." *New York Times*, May 22 (https://www.nytimes.com/2019/05/22/travel/unlikely-hikers-hit-the -trail.html).

Kruesi, Kimberlee. 2017. "Lawmakers Strip Climate Change References from New Idaho K–12 Science Standards." *Idaho Statesman*, February 7 (http://www.idahostatesman.com/news/politics-government/state-politics /article131754369.html#storylink=cpy).

LaDuke, Winona. 2016. *The Winona Laduke Chronicles: Stories from the Front Lines in the Battle for Environmental Justice*, edited by Aaron Cruz. Ponsford, MN, and Winnipeg, MB: Spotted Horse Press and Fernwood Publishing.

———. 2019. "We Are the Home Team, We Are Water Protectors." Honor the Earth. Retrieved August 6, 2021 (https://www.honortheearthmerchandise .com/blog/2019/12/2/19-yearendcelebration).

———. 2020. *To Be a Water Protector: The Rise of the Wiindigoo Slayers.* Blackpoint, NS, and Ponsford, MN: Fernwood Publishing and Spotted Horse Press.

Lameman, Crystal. 2014. "Kihci Pikiskwewin: Speaking the Truth." Pp. 118–26 in *A Line in the Tar Sands: Struggles for Environmental Justice,* edited by Toban Black, Stephen D'Arcy, Tony Weis, and Joshua Kahn Russell. Oakland, CA: PM Press.

Land Report editors. 2015. "2015 Land Report 100." *Land Report,* October 15 (https://landreport.com/2015/10/2015-land-report-100/).

LeQuesne, Theo. 2019. "From Carbon Democracy to Carbon Rebellion: Countering Petro-Hegemony on the Frontlines of Climate Justice." *Journal of World-Systems Research* 25(1):16–27.

Lewis, Renee. 2015. "As UN Says World to Warm by 3 Degrees, Scientists Explain What That Means." *Aljazeera America,* September 23 (http:// america.aljazeera.com/articles/2015/9/23/climate-change-effects-from-a-3-c -world.html).

Lippmann, Walter. 1922. *Public Opinion.* New York: Harcourt, Brace and Co. (http://wwnorton.com/college/history/america-essential-learning/docs /WLippmann-Public_Opinion-1922.pdf).

Lipsitz, George. 2006. "Unexpected Affiliations: Environmental Justice and the New Social Movements." *Work and Days* 24(47/48):25–44.

Loan Trần, Ngọc 2013. "Calling IN: A Less Disposable Way of Holding Each Other Accountable." *Black Girls Dangerous.* Retrieved August 20, 2016 (http://www.blackgirldangerous.org/2013/12/calling-less-disposable-way -holding-accountable/).

Loomis, Erik. 2015. *Out of Sight: The Long and Disturbing Story of Corporations Outsourcing Catastrophe.* New York: The New Press.

Lucas, Anne E. 2004. "No Remedy for the Inuit: Accountability for Environmental Harms under U.S. and International Law." Pp. 191–206 in *New Perspectives on Environmental Justice: Gender, Sexuality, and Activism,* edited by Rachel Stein. New Brunswick, NJ: Rutgers University Press.

Malin, Stephanie A., and Kathryn Teigen DeMaster. 2016. "A Devil's Bargain: Rural Environmental Injustices and Hydraulic Fracturing on Pennsylvania's Farms." *Journal of Rural Studies* 47, Part A:1–13.

Malloy, Chuck. 2015. "Idaho Might Soon Be Cooking with Gas." *Idaho Press,* September 10 (https://www.idahopress.com/members/idaho-may-soon-be -cooking-with-gas/article_04a05544-5751-11e5-9ceb-3bf34f07d5d6.html).

Mammoet Canada Western Ltd. 2009. "Imperial Oil Limited Lewiston, ID to Kearl Lake Project, AB. Validation Module Route Study Execution Plan. Project No.: 0010030265-P042" (http://media.spokesman.com/documents /2011/04/Validation_Module_Route_Study_preface.pdf).

Marlon, Jennifer, Peter Howe, Matto Mildenberger, Anthony Leiserowitz, and Xinran Wang. 2018. "Yale Climate Opinion Maps 2018." Yale Program on Climate Change Communication. Retrieved January 22, 2019 (http://

climatecommunication.yale.edu/visualizations-data/ycom-us-2018/?est= happening&type=value&geo=county).

Martínez, María. 2018. "Reiteraciones Relacionales y Activaciones Emociona-les: Hacia una Radicalización de la Procesualidad de Las Identidades Col-ectivas a través del Estudio de Las Movilizaciones Feministas en el Estado Español." *Athenea Digital: Revista de Pensamiento e Investigación Social* 18(1):293–317.

Marx, Karl. [1844] 1978. "Contribution to the Critique of Hegel's *Philosophy of Right*: Introduction." Pp. 53–65 in *The Marx-Engels Reader*, edited by Robert C. Tucker. New York: W.W. Norton & Company.

———. [1888] 1978. "Theses on Feuerbach." Pp. 143–45 in *The Marx-Engels Reader*, edited by Robert C. Tucker. New York: W.W. Norton & Company.

McAdam, Doug. 1986. "Recruitment to High-Risk Activism: The Case of Free-dom Summer." *American Journal of Sociology* 92(1):64–90.

———. 1988. *Freedom Summer*. New York: Oxford University Press.

McCright, Aaron M., Sandra T. Marquart-Pyatt, Rachael L. Shwom, Ste-ven R. Brechin, and Summer Allen. 2016. "Ideology, Capitalism, and Cli-mate: Explaining Public Views about Climate Change in the United States." *Energy Research & Social Science* 21:180–89.

McKenzie, Lisa M., Ruixin Guo, Roxana Z. Witter, David A. Savitz, Lee S. Newman, and John. L. Adgate. 2014. "Birth Outcomes and Maternal Resi-dential Proximity to Natural Gas Development in Rural Colorado." *Envi-ronmental Health Perspectives* 112(4):412–17.

McKibben, Bill. 2012. "Global Warming's Terrifying New Math." *Rolling Stone*, July 19 (http://www.rollingstone.com/politics/news/global-warmings -terrifying-new-math-20120719).

———. 2016a. "Global Warming's Terrifying New Chemistry." *The Nation*, March 23 (http://www.thenation.com/article/global-warming-terrifying-new -chemistry/).

———. 2016b. "A World at War." *The New Republic*, August 15 (https:// newrepublic.com/article/135684/declare-war-climate-change-mobilize-wwii ?utm=3500rg).

McLachlan, Stéphane M. 2014. "'Water Is a Living Thing' Environmental and Human Health Implications of the Athabasca Oil Sands for the Mikisew Cree First Nation and Athabasca Chipewyan First Nation in Northern Alberta." Mikisew Cree First Nation and Athabasca Chipewyan First Nation.

McLeod, John D. 1993. "The Search for Oil and Gas in Idaho." *GeoNote* vol. 21. Moscow, ID: Idaho Geologic Survey (https://www.idahogeology.org /product/g-21).

McPherson, Miller, Lynn Smith-Lovin, and James M Cook. 2001. "Birds of a Feather: Homophily in Social Networks." *Annual Review of Sociology* 27(1):415–44.

Megerian, Chris. 2015. "China and the World Turn to California for Climate Change Expertise." *Los Angeles Times*, December 7 (http://www.latimes .com/politics/la-me-pol-sac-climate-california-china-20151207-story.html).

Melucci, Alberto. 1989. *Nomads of the Present: Social Movements and Indi-vidual Needs in Contemporary Society*. London: Hutchinson Radius.

Mills, C. Wright. 1959. *The Sociological Imagination*. New York: Oxford University Press.

Moe, Kristin. 2012. "Alberta Tar Sands Illegal under Treaty 8, First Nations Charge." *Truthout*, October 19 (https://truthout.org/articles/alberta-tar-sands-illegal-under-treaty-8-first-nations-charge/).

Molotch, Harvey. 1970. "Oil in Santa Barbara and Power in America." *Sociological Inquiry* 40:131–44.

Montrie, Chad. 2003. *To Save the Land and People: A History of Opposition to Surface Mining in Appalachia*. Chapel Hill: University of North Carolina Press.

Moore, Hilary, and Joshua Kahn Russell. 2011. *Organizing Cools the Planet: Tools and Reflections to Navigate the Climate Crisis*. Oakland, CA: PM Press.

Müller, Melanie, and Heike Walk. 2014. "Democratizing the Climate Negotiations System through Improved Opportunities for Participation." Pp. 31–43 in *Routledge Handbook of the Climate Change Movement*, edited by Matthias Dietz, and Heiko Garrelts. New York: Routledge.

Naples, Nancy A., ed. 1998. *Community Activism and Feminist Politics: Organizing Across Race, Class, and Gender*. New York: Routledge.

NASA. 2021. "2020 Tied for Warmest Year on Record, NASA Analysis Shows." Retrieved June 1, 2021 (https://www.nasa.gov/press-release/2020-tied-for-warmest-year-on-record-nasa-analysis-shows).

Nellis, Stephen. 2014. "$114M Slated to Fuel Santa Maria Energy Operations." *Santa Maria Business Times*, January 10 (https://www.pacbiztimes.com/2014/01/10/114m-slated-to-fuel-santa-maria-energy-operations/).

Neuhauser, Alan. 2016. "Shell, ConocoPhillips Drop Arctic Drilling Plans." *U.S. News*, May 10 (https://www.usnews.com/news/articles/2016-05-10/shell-conocophillips-drop-arctic-drilling-plans).

Nez Perce Tribal Executive Committee. 2016. "Re: Request for Comments on Federal Infrastructure Project Decision Making." November 30 (https://www.bia.gov/sites/bia.gov/files/assets/as-ia/raca/pdf/idc2-055636.pdf).

Nixon, Rob. 2011. "Introduction." Pp. 1–44 in *Slow Violence and the Environmentalism of the Poor*. Cambridge, MA: Harvard University Press.

Ogburn, Stephani. 2011. "Idaho: The CAFO State?" *High Country News*, August 22 (http://www.hcn.org/issues/43.14/idaho-the-cafo-state/#comments).

OPEC. 2017. "Brief History." Retrieved April 17, 2017 (http://www.opec.org/opec_web/en/about_us/24.htm).

Oreskes, Naomi, and Erik M. Conway. 2010. *Merchants of Doubt: How a Handful of Scientists Obscured the Truth on Issues from Tobacco Smoke to Global Warming*. New York: Bloomsbury Press.

Pellow, David N. 2016. "Environmental Justice and Rural Studies: A Critical Conversation and Invitation to Collaboration." *Journal of Rural Studies* 47, Part A:381–86.

Pellow, David Naguib. 2014. *Total Liberation: The Power and Promise of the Animal Rights and Radical Earth Movement*. Minneapolis: University of Minnesota Press.

The People's Climate Movement. 2017. "We Resist. We Build. We Rise." Retrieved February 6, 2017 (https://peoplesclimate.org/?source=350).

Perks, Rob. 2009. "Appalachian Hearthbreak: Time to End Mountaintop Removal Coal Mining." National Resources Defense Council. November 9 (http://www.nrdc.org/land/appalachian/files/appalachian.pdf).

Pitt, Hannah, Kate Larsen, and Maggie Young. 2020. "The Undoing of US Climate Policy: The Emissions Impact of Trump-Era Rollbacks." Rhodium Group. September 17 (https://rhg.com/research/the-rollback-of-us-climate -policy/).

Plumer, Brad. 2014. "Obama Says Fracking Can Be a 'Bridge' to a Clean-Energy Future. It's Not That Simple." *Washington Post*, January 29 (https://www .washingtonpost.com/news/wonk/wp/2014/01/29/obama-says-fracking -offers-a-bridge-to-a-clean-energy-future-its-not-that-simple/?utm_term= .4323fc9aacf1).

Polletta, Francesca. 2002. *Freedom Is an Endless Meeting: Democracy in American Social Movements*. Chicago: University of Chicago Press.

———. 2006. *It Was Like a Fever: Storytelling in Protest and Politics*. Chicago: University of Chicago Press.

Popovich, Nadja, Livia Albeck-Ripka, and Kendra Pierre-Louis. 2021. "The Trump Administration Rolled Back More Than 100 Environmental Rules. Here's the Full List." *New York Times*, January 20.

Prentice, George. 2011a. "Fracking Idaho." *Pacific Northwest Inlander*, May 25 (http://www.inlander.com/spokane/fracking-idaho/Content?oid=2135066& mode=print).

———. 2011b. "Bridge under Troubled Waters: Payette County Natural Gas Company Faces Uncertain Future." *Boise Weekly*, October 5.

———. 2014. "Update: Idaho Anti-Fracking Activists Released after 7-Day Incarceration." *Boise Weekly*, October 17.

Princeton Review. 2015. "Guide to 353 Green Colleges: 2015 Edition Press Release" (http://www.princetonreview.com/press/green-guide/press-release).

———. 2021. "Top 50 Green Colleges." Retrieved July 26, 2021 (https://www .princetonreview.com/college-rankings?rankings=top-50-green-colleges& ceid=green-colleges).

Pulido, Laura. 1996. "A Critical Review of the Methodology of Environmental Racism Research." *Antipode* 28(2):142–59.

Radetzki, Marian, and Roberto F. Auilera. 2016. "The Age of Cheap Oil and Natural Gas Is Just Beginning." *Scientific American*, May 3 (https://blogs .scientificamerican.com/guest-blog/the-age-of-cheap-oil-and-natural-gas-is -just-beginning/).

Reed, Jean-Pierre, and John Foran. 2002. "Political Cultures of Opposition." *Critical Sociology* 28(3):335–70.

Retail Industry Leaders Association, Information Technology Industry Council, and Clean Edge. 2017. *Corporate Clean Energy Procurement Index: State Leadership and Rankings*. Clean Edge (http://cleanedge.com/reports /Corporate-Clean-Energy-Procurement-Index).

Reyes, Oscar. 2012. "What Goes Up Must Come Down—Carbon Trading, Industrial Subsidies and Capital Market Governance." Pp. 185–209, in *What Next: Climate, Development and Equity*, edited by Niclas Hällström. Uppsala, Sweden: What Next Forum and Dag Hammarskjöld Foundation.

Rosenthal, Elisabeth. 2012. "Race Is On as Ice Melt Reveals Arctic Treasures." *New York Times*, September 18 (http://www.nytimes.com/2012/09 /19/science/earth/arctic-resources-exposed-by-warming-set-off-competition .html?ref=greenland).

Rosewarne, Stuart, James Goodman, and Rebecca Pearse. 2014. *Climate Action Upsurge: The Ethnography of Climate Movement Politics*. London: Routledge.

Russell, Joshua Kahn, Linda Capato, Matt Leonard, and Rae Breaux. 2014. "Lessons from Direct Action at the White House to Stop the Keystone XL Pipeline." Pp. 166–80 in *A Line in the Tar Sands: Struggles for Environmental Justice*, edited by Toban Black, Stephen D'Arcy, Tony Weis, and Joshua Kahn Russell. Oakland, CA: PM Press.

Saad, Lydia, and Jeffrey M. Jones. 2016. "U.S. Concern about Global Warming at Eight-Year High." Gallup. March 16 (http://www.gallup.com/poll /190010/concern-global-warming-eight-year-high.aspx).

Sahagun, Louis. 2014. "U.S. Officials Cut Estimate of Recoverable Monterey Shale Oil by 96%." *Los Angeles Times*, May 20 (http://www.latimes.com /business/la-fi-oil-20140521-story.html).

Sampson, Anthony. 1975. *The Seven Sisters: The Great Oil Companies and the World They Made*. New York: Viking Press.

Sartre, Jean-Paul. 1992. *The Age of Reason: A Novel*. New York: Vintage Books.

Saxifrage, Barry. 2013. "Climate Pollution: 140 Nations vs. Alberta's Tar Sands." Visual Carbon. Retrieved July 26, 2021 (http://www.saxifrages.org /eco/go58h/Climate_pollution_140_nations_vs_Albertas_tar_sands).

Scott, Rebecca R. 2010. *Removing Mountains: Extracting Nature and Identity in the Appalachian Coalfields*. Minneapolis: University of Minnesota Press.

Seager, Joni. 1996. "'Hysterical Housewives' and Other Mad Women." Pp. 271–83 in *Feminist Political Ecology: Global Issues and Local Experiences*, edited by Dianne Rocheleau, Barbara Thomas-Slayter, and Esther Wangari. London: Routledge.

Sedbrook, Danielle. 2016. "2,4-D: The Most Dangerous Pesticide You've Never Heard Of." National Resources Defense Council. Retrieved December 20, 2016 (https://www.nrdc.org/stories/24-d-most-dangerous-pesticide-youve -never-heard).

Shabecoff, Phil. 1988. "Global Warming Has Begun, Expert Tells Senate." *New York Times*, June 24 (http://www.nytimes.com/1988/06/24/us/global -warming-has-begun-expert-tells-senate.html?pagewanted=all).

Siegler, Kirk. 2017. "Leaving Urban Areas for the Political Homogeniety of Rural Towns." NPR, February 14 (http://www.npr.org/2017/02/14/512875545 /leaving-urban-areas-for-the-political-homogeneity-of-rural-towns).

Silk, Ezra. 2016. *Victory Plan*. The Climate Mobilization. https://www.the climatemobilization.org/wp-content/uploads/2020/07/Victory-Plan-July -2020-Update.pdf).

Simpson, Mike. 2010. "Simpson Supports Domestic Energy Production." Retrieved March 12, 2016 (http://simpson.house.gov/news/documentsingle .aspx?DocumentID=177081).

Smith, Andrea. 2008. *Native Americans and the Christian Right: The Gendered Politics of Unlikely Alliances.* Durham, NC: Duke University Press.

Smith, Brian. 2014a. "Changes to Idaho's Oil and Gas Industry Proposed." MagicValley.com, August 2 (http://magicvalley.com/news/local/changes-to-idaho-s-oil-and-gas-industry-proposed/article_915089b0-19da-11e4-9ea6-0019bb2963f4.html).

———. 2014b. "Conservationists Alarmed as Gas Explorer's Finances Downgraded." MagicValley.com, August 21 (http://magicvalley.com/business/conservationists-alarmed-as-gas-explorer-s-finances-downgraded/article_1cbccaf2-28f1-11e4-880f-0019bb2963f4.html).

Smith, Jackie, and Nicole Doerr. 2016. "Democratic Innovation in the U.S. and European Social Forums." Pp. 339–59 in *Handbook of the World Social Forums*, edited by Jackie Smith, Scott Byrd, Ellen Reese, and Elizabeth Smythe. New York: Routledge.

Snow, David. 2004. "Framing Processes, Ideology, and Discursive Fields." Pp. 380–412 in *The Blackwell Companion to Social Movements*, edited by David A. Snow, Sarah A. Soule, and Hanspeter Kriesi. Malden, MA: Blackwell.

Snow, David A., E. Burke Rochford, Steven K. Worden, and Robert D. Benford. 1986. "Frame Alignment Processes, Micromobilization, and Movement Participation." *American Sociological Review* 51(4):464–81.

Snow, David A., Sarah A. Soule, and Hanspeter Kriesi. 2007. *The Blackwell Companion to Social Movements.* Malden, MA: Blackwell.

Solnit, Rebecca. 2016. *Hope in the Dark: Untold Histories, Wild Possibilities.* Chicago: Haymarket Books.

Southwest Network for Environmental and Economic Justice. 1996. "Jemez Principles for Democratic Organizing" (https://www.ejnet.org/ej/jemez.pdf).

Sovacool, Benjamin K. 2014. "What Are We Doing Here? Analyzing Fifteen Years of Energy Scholarship and Proposing a Social Science Research Agenda." *Energy Research & Social Science* 1:1–29.

The State Energy and Environmental Impact Center. 2019. "Climate and Health Showdown in the Courts: State Attorneys General Prepare to Fight." NYU School of Law (https://www.law.nyu.edu/sites/default/files/climate-and-health-showdown-in-the-courts.pdf).

Stein, Rachel, ed. 2004. *New Perspectives on Environmental Justice: Gender, Sexuality, and Activism.* New Brunswick, NJ: Rutgers University Press.

Stephenson, Wen. 2015. *What We Are Fighting For Now Is Each Other: Dispatches from the Front Lines of Climate Justice.* Boston, MA: Beacon Press.

Strauss, Benjamin. 2015. "Coastal Nations, Megacities Face 20 Feet of Sea Rise." Climate Central (http://www.climatecentral.org/news/nations-megacities-face-20-feet-of-sea-level-rise-19217).

Summars, Emily. 2015. "Who's at Fault? Quake Calls Defined by Insurance Policies." *Enid News & Eagle*, October 4 (http://www.enidnews.com/news/earthquakes/who-s-at-fault-quake-calls-defined-by-insurance-policies/article_9f716a82-6a4c-11e5-8doc-ob66c3623coe.html).

Taliman, Valerie. 2016. "Veterans Ask Forgiveness and Healing in Standing Rock." *Indian Country Today*, December 7.

Taylor, Dorceta E. 2016. *The Rise of the American Conservation Movement: Power, Privilege, and Environmental Protection.* Durham, NC: Duke University Press.

Taylor, Verta, and Nancy Whittier. 1992. "Collective Identity in Social Movement Communitites: Lesbian Feminist Mobilization." Pp. 104–29 in *Frontiers in Social Movement Theory*, edited by Aldon D. Morris, and Carol Mueller. New Haven, CT: Yale University Press.

Thomas-Muller, Clayton. 2014. "The Rise of the Native Rights–Based Strategic Framework." Pp. 240–52 in *A Line in the Tar Sands: Struggles for Environmental Justice*, edited by Toban Black, Stephen D'Arcy, Tony Weis, and Joshua Kahn Russell. Oakland, CA: PM Press.

350 Santa Barbara. "About Us." Retrieved February 5, 2019 (https://350sb.org /about/).

350.org Staff. 2016. "Orlando." Retrieved February 18, 2018 (http://350.org /orlando/?akid=14359.599245.kVHD7h&rd=1&t=9).

350.org and Attach France, eds. 2015. *Crime Climatique Stop!* Paris: Le Seuil (https://350.org/climate-crimes/).

Tsing, Anna Lowenhaupt. 2004. *Friction: An Ethnography of Global Connection.* Princeton, NJ: Princeton University Press.

Turnbull, David. 2016. "The Not-So-Hidden Fracking Money Fueling the 2016 Elections." *Oil Change International.* Retrieved March 12, 2016 (http:// priceofoil.org/2016/02/09/the-not-so-hidden-fracking-money-fueling-the -2016-elections/).

Turner, A. J., D. J. Jacob, J. Benmergui, S. C. Wofsy, J. D. Maasakkers, A. Butz, O. Hasekamp, and S. C. Biraud. 2016. "A Large Increase in U.S. Methane Emissions over the Past Decade Inferred from Satellite Data and Surface Observations." *Geophysical Research Letters* 43(5):2218–24.

US Census Bureau. 2019a. "QuickFacts." Retrieved August 7, 2021 (https:// www.census.gov/quickfacts/fact/table/US/PST045219).

———. 2019b. "Small Area Income and Poverty Estimates." Retrieved August 7, 2021 (https://www.census.gov/data-tools/demo/saipe/#/?map_geo Selector=aa_c).

U.S. Energy Information Administration. 2016a. "United States Remains Largest Producer of Petroleum and Natural Gas Hydrocarbons." International Energy Agency. Retrieved February 17, 2017 (http://www.eia.gov /todayinenergy/detail.php?id=26352).

———. 2016b. "Crude Oil Production." Retrieved September 1, 2016 (http:// www.eia.gov/dnav/pet/pet_crd_crpdn_adc_mbbl_a.htm).

———. 2016c. "Hydraulically Fractured Wells Provide Two Thirds of U.S. Natural Gas Production" (http://www.eia.gov/todayinenergy/detail.php?id=26112).

———. 2016d. "Hydraulic Fracturing Accounts for About Half of Current U.S. Crude Oil Production" (http://www.eia.gov/todayinenergy/detail.php ?id=25372).

———. 2017. "Drilling Productivity Report." Retrieved April 18, 2017 (https:// www.eia.gov/petroleum/drilling/#tabs-summary-2).

———. 2021a. "Drilling Productivity Report." Retrieved July 26, 2021 (https:// www.eia.gov/petroleum/drilling/#tabs-summary-2).

———. 2021b. "Crude Oil Production." Retrieved July 29, 2021 (http://www
.eia.gov/dnav/pet/pet_crd_crpdn_adc_mbbl_a.htm).

US EPA. 2011. "EPA Issues Final Guidance to Protect Water Quality in Appalachian Communities from Impacts of Mountaintop Mining." Water Online, July 26. Retrieved April 18, 2017 (https://www.wateronline.com/doc/epa-issues-final-guidance-to-protect-water-0001).

———. 2015. "J.R. Simplot Company Clean Air Act (CAA) Settlement." Retrieved March 12, 2016 (https://www.epa.gov/enforcement/jr-simplot-company-clean-air-act-caa-settlement).

———. 2016. "Hydraulic Fracturing for Oil and Gas: Impacts from the Hydraulic Fracturing Water Cycle on Drinking Water Resources in the United States." Washington DC: Office of Research and Development (https://cfpub.epa.gov/ncea/hfstudy/recordisplay.cfm?deid=332990).

———. 2017. "Environmental Justice." Retrieved February 13, 2017 (https://www.epa.gov/environmentaljustice).

UCSB Sustainability. 2019. "UCSB Sustainability." Retrieved Feburary 27, 2019 (http://www.sustainability.ucsb.edu/ucsb-sustainability/).

Union of Concerned Scientists. 2007. *Smoke, Mirrors and Hot Air: How ExxonMobil Uses Big Tobacco's Tactics to Manufacture Uncertainty on Climate Science* (https://web.archive.org/web/20150726204316/http://www.ucsusa.org/sites/default/files/legacy/assets/documents/global_warming/exxon_report.pdf).

———. 2013. "What Are Tar Sands?" Retrieved January 27, 2022 (http://www.ucsusa.org/clean-vehicles/all-about-oil/what-are-tar-sands#.WLh3dhLyvaY).

———. 2015. *The Climate Deception Dossiers: Internal Fossil Fuel Industry Memos Reveal Decades of Corporate Disinformation* (http://www.ucsusa.org/sites/default/files/attach/2015/07/The-Climate-Deception-Dossiers.pdf).

United Nations. 1992. *United Nations Framework Convention on Climate Change.* Rio de Janeiro.

Urban Indian Health Institute. 2019. *Missing and Murdered Indigenous Women and Girls.* Seattle: Indian Health Board.

Urbina, Ian. 2011. "Rush to Drill for Natural Gas Creates Conflicts with Mortgages." *New York Times*, October 19 (http://www.nytimes.com/2011/10/20/us/rush-to-drill-for-gas-creates-mortgage-conflicts.html?_r=0).

Urry, John. 2011. *Climate Change and Society.* Cambridge: Polity Press.

USGS. 2008. "Selected Radiochemical and Chemical Constituents and Physical Properties of Water in the Snake River Plain Aquifer." *Scientific Investigations Report.* US Department of the Interior (https://pubs.usgs.gov/sir/2008/5089/section8.html).

Vandehey, Scott. 2013. "Local Power: Harnessing NIMBYism for Sustainable Suburban Energy Production." Pp. 242–55 in *Cultures of Energy: Power, Practices, Technologies*, edited by Sarah Strauss, Stephanie Rupp, and Thomas Love. Walnut Creek, CA: Left Coast Press.

Van Dyke, Nella, and Holly J. McCammon, eds. 2010. *Strategic Alliances: Coalition Building and Social Movements.* Minneapolis: University of Minnesota Press.

Voyles, Traci Brynne. 2015. *Wastelanding: Legacies of Uranium Mining in Navajo Country*. Minneapolis: University of Minnesota Press.

Warner, Mark R., and Tim M. Kaine. 2014. "Letter to Anthony Foxx, Secretary of Transportation." Washington, DC: United States Senate (https://www.kaine.senate.gov/press-releases/sens-warner-and-kaine-call-for-improved-crude-oil-rail-safety-better-communication-with-first-responders-following-lynchburg-train-derailment).

Weiner, Jon. 2016. "Median Installed Price of Solar in the United States Fell by 5–12% in 2015." Berkeley Lab, August 24 (http://newscenter.lbl.gov/2016/08/24/median-installed-price-solar-united-states-fell-5-12-2015/).

Whyte, Kyle Powys. 2017. "The Dakota Access Pipeline, Environmental Injustice, and U.S. Colonialism." *Red Ink: An International Journal of Indigenous Literature, Arts, & Humanities* (19.1):154–69.

Wild Idaho Rising Tide. 2015. "Oil and Coal Train Update from Wild Idaho Rising Tide." *Earth First! Journal*. Retrieved: April 14, 2015 (http://earthfirstjournal.org/newswire/2015/01/15/oil-and-coal-train-update-from-wild-idaho-rising-tide/).

Williams, Emily, and Theo LeQuesne. 2019. "The University of California Has Finally Divested from Fossil Fuels." *The Nation*, October 8 (https://www.thenation.com/article/archive/california-fossil-fuels/).

Williams, Raymond. 1960. *Culture and Society: 1780–1950*. London: Chatto and Windus.

Willow, Anna. 2019. *Understanding ExtrACTIVISM: Culture and Power in Natural Resource Disputes*. New York: Routledge.

The Workforce Investment Board of Santa Barbara. 2012. "Industry, Employment & Skills in a Time of Transition: An Employment Forecast for Santa Barbara County" (http://www.sbcwdb.org/uploadedFiles/sbcwdb/content-2020/reports/workforce-intelligence/WIB-2012-Employment-Forecast-Presentation.pdf).

World Resources Institute. 2014. "6 Graphs Explain the World's Top 10 Emitters." Retrieved February 27 2019 (http://www.wri.org/blog/2014/11/6-graphs-explain-world%E2%80%99s-top-10-emitters).

Yardley, William. 2010. "Talk of Selling an Empty House Has Boise Buzzing." *New York Times*, February 15 (http://www.nytimes.com/2010/02/16/us/16idaho.html).

Zammit-Lucia, Joseph. 2013. "COP19: The UN's Climate Talks Proved to Be Just Another Cop Out." *The Guardian*, December 2 (https://www.theguardian.com/sustainable-business/cop19-un-climate-talks-another-cop-out).

Zeller, Tom Jr. 2010. "Oil Sands Effort Turns on a Fight over a Road." *New York Times*, October 21 (https://www.nytimes.com/2010/10/22/business/energy-environment/22road.html).

Index

Notes: Figures are noted as *fig* and tables are noted as *tab*.

Founded in 1893,
UNIVERSITY OF CALIFORNIA PRESS
publishes bold, progressive books and journals
on topics in the arts, humanities, social sciences,
and natural sciences—with a focus on social
justice issues—that inspire thought and action
among readers worldwide.

The UC PRESS FOUNDATION
raises funds to uphold the press's vital role
as an independent, nonprofit publisher, and
receives philanthropic support from a wide
range of individuals and institutions—and from
committed readers like you. To learn more, visit
ucpress.edu/supportus.